The Nature of Our Cities

The Nature
of Our Cities

HARNESSING THE POWER
OF THE NATURAL WORLD TO
SURVIVE A CHANGING PLANET

Nadina Galle

MARINER BOOKS

New York   Boston

HarperCollins books may be purchased for educational,
business, or sales promotional use. For information, please
email the Special Markets Department at
SPsales@harpercollins.com.

FIRST EDITION

Designed by Chloe Foster

Library of Congress Cataloging-in-Publication Data
has been applied for.

ISBN 978-0-06-332261-5

24 25 26 27 28 LBC 5 4 3 2 1

For Luca, for being with me,
every step of the way.

CONTENTS

A City Without Nature

A BEAD OF SWEAT TRICKLING down the side of your face confirms that you're still awake. You get up to peel your sweat-soaked sheets from the bed, removing the fresh pair you'd washed from the dryer, knowing this would be another blistering night filled with everything but sleep. As you tuck each corner of the new sheets around your bed, you remind yourself that you've survived worse.

The scream of a skidding car outside ends in a shattering thud. You've heard that sound before—last June, when temperatures got so high that oil, built up from millions of tires scraping the road, seemed to rise from the pores in the asphalt. You peek through your curtains to find that the passengers have climbed safely out of their battered vehicles, and you dutifully call in the crash. It's day eight of what's expected to be another weeks-long heat wave, and each day has been worse than the last.

Then, you remember your ninety-year-old neighbor. Did she make it through the night?

It takes twelve knocks, but finally, she answers her door.

"What's the hurry?" she says, and you give her the longest, sweatiest hug you've ever given someone.

You hop in the shower. The lukewarm water reminds you that the power has been out since yesterday, the grid overloaded by people trying to keep cool with air conditioners. With the tram down, its cables melting, you grab your bike from the basement. The gears skip, rusty from last year's flood, but you've learned how to work with them. With a soft pedal and an even push on the shift, you ease it into the right gear and make your way to work. You'd started the summer biking every day, but after a truck knocked you to the pavement last month, you haven't mustered up the courage to climb back into the saddle. With the office too far to walk to, you have no choice today.

Pedaling down Main Street, your eyes sting from the smog—the viscous air sags in your path, it's like the pollution adds wind resistance. Then you remember that your asthma always acts up in the summer. The smoke rolling in from the wildfire hundreds of miles to the west isn't helping. Asthma had started ten years ago with a familiar itch in your throat, then a persistent cough. Your doctor called it "the Beijing cough." The name struck you as strange, with millions dying of air pollution in cities worldwide.

You turn right onto a small avenue, one of the few where large, mature pin oaks tower over the neighborhood. As you pedal under their canopy, your throat relaxes. You stop and take a couple of gulps of the water you brought.

Leaning against one of the trunks, you swear you can feel it teeter away from you. Stunned, you step back and notice the oak's bottom branches for the first time—they're drooping, their leaves browning.

"Won't take long for these to come down," a man's voice calls behind you. He wipes sweat from his brow, stopping to let his dog out on a small speckle of green. Even without a leash, the dog paces in a tight circle, like it has forgotten how to play. You wave and pedal on.

You pull up to the office, the first floor of a small wooden 1970s structure. Because of the era in which it was built, before the flood maps got redrawn, it sits lower than its neighbors. The emergency generator installed five heat waves ago means the lights are on. But the AC remains broken from the flood last year. The brown mold on the walls reminds you just how high the water got. You look away before your brain replays the trauma of that day, but you know it's too late to block the memory.

Until you saw those walls again, you would have struggled to imagine this place filled with floodwater. The looming threat of wildfire, blared every day by meteorologists, makes such an abundance of water seem almost aspirational—between this August's extreme heat and drought, you can see the city is a tinderbox. It only took one spark to burn a neighboring town to the ground last week. You wince, remembering Ellis, lost to the blaze.

Emergency counselors tried to support the families and friends of victims. But you found their advice of little help: "If you feel yourself getting anxious about the fires," the kind woman had told you, "take deep, calming breaths." When you tried it yourself, the smoky air made you gasp.

You leave work early in the afternoon to escape the hottest part of the day, stopping at a small park. The last meadow left in a city built atop them includes the only stream that hasn't been routed into an underground pipe. As a result, you and too many of your hundreds of thousands of neighbors are there, demanding more from these four acres than they can provide. The park resembles Woodstock, right before Jimi Hendrix came on, though the air has a tinge that your Hippie parents would not have recognized.

Dragging your bike behind you, you elbow through the crowd and find solace behind a dried-up worm hotel, another failed attempt to engage the community in composting. You sit cross-legged, inhaling the musk of that dusty air, trying to take those deep, supposedly calming breaths. But as the sweat once again pools on your furrowed brow—you worry the world has changed forever.

CHAPTER 1

Unsilenced Spring

I believe in God, only I spell it Nature.
—Frank Lloyd Wright, American architect

THE NEIGHBORHOOD WAS DYING slowly—that much everyone could agree on. The nightmare had started with a celebration on October 10, 1959, when the city opened a new four-lane highway that traversed Maastricht, a college town on the southern tip of the Netherlands. Neighbors gathered along the concrete walls of the new thoroughfare, admiring the perfectly smooth lanes, feeling the vibration of the first speeding cars thunder through their bodies. The city had promised that the new highway, with its 100-kilometer-per-hour speed limits, would usher in a new era of mobility for the city.

The reality was far from it. The imposing steel highway, an extension of the one that stretched across the Netherlands, quickly became a source of misery for adjacent residents. The city tore down blocks to make way for the arterial road, uprooting families and the street trees they tended. The absence of these trees, whose leaves filter and purify the air, only exacerbated the cities' worsening air and noise pollution. The residents who were lucky enough not to lose their homes to the widened footprint of the highway found their conversations drowned out by the roar of the 50,000 engines passing every day. The day that the construction started, their onetime neighbors suddenly became inaccessible despite living just two blocks away, separated by walls of thumping concrete.

Soon, asthma levels began to spike. Then, within a few decades, the thoroughfare that had promised zippy commutes was packed with bumper-to-bumper traffic. Residents along the route began to notice that their parked cars and porches were coated in gray dust, which turned out to be the carcinogenic by-product of millions of drivers pumping their brakes.

In just a few years, what had once been the neighborhood's meeting place—the site of evening barbecues and pickup basketball games—had become a hazard. Crossing from one side to the other was a matter of life and death.

And yet, it still was not enough highway. The country's leaders saw the Maastricht stretch of A2 as a bottleneck, delaying movement between the populous Dutch coastline and its neighbors to the south. They proposed expanding the highway to ease the traffic that had become crippling. Thankfully it was a bridge too far.

Maastricht's elders, who had seen how such promises played out, resisted, rallying residents across the city to demand a better solution. Through decades of activism, they built a coalition between local governments and the head of Public Works and Water Management. Instead of expanding the highway—literally widening the city's divide— the alliance agreed to take a different approach. It would tunnel the motorway underground, separating local and (inter)national traffic into a stack of lanes to ease slowdowns and transform the surface into a meandering park. Lined with enormous trees, rolling flower meadows, and winding pedestrian and bike paths, the park would help mitigate the noise, air pollution, and physical barriers that decades of bigger and bigger roads had unleashed on the city.

Today, the park is known as the Groene Loper, or the Green Carpet. Its trees provide varied habitats for the area's resurgent biodiversity, homes for the pipistrelle bat, the common swift, the house sparrow, and hundreds of species of insects, many of which pollinate the park's flowers. Road safety has improved considerably, with significant reductions in noise pollution and vastly enhanced air quality. Maastricht met the country's legal air-quality standards for the first time in seven decades. Surrounding communities, whose members sometimes had to walk twice as far to reach their nearest bakery, can now connect while gardening, walking, and playing in their newfound oasis.

"I used to live along a highway," one resident told the local news station. "Now I live along a park. And I didn't even have to move."

Building from research into the impact of urban trees on the health of their human neighbors, the massive construction project had one particularly ambitious goal: to increase the life expectancy of the residents surrounding it by five years. To make that happen, its newly planted 1,800 trees needed to thrive.

Tjeu Franssen (whose name was changed for privacy reasons) was tasked with nurturing this new urban forest. An arborist by trade, he accepted the heavy burden because he believed in the mission and valued the ability to work outside, away from a computer, the quality that had attracted him to his career nearly thirty years ago.

I met Tjeu one morning on one of the excursions I'd take around Maastricht to water its neighborhood trees. It was a weekend ritual I enjoyed, an informal way to learn about other Dutch cities while I did research for my Ph.D. in ecological engineering. Wanting to get to know the steward of my favorite park in the city, I eventually mustered the courage to reach out to him and invite him along.

As we strolled down a sunny block in the city's historic Jekerkwartier district, Tjeu stopped at a small beech tree that tilted over the road as if exhausted. He grabbed one of its dry leaves with a fist, almost crunching it to powder. "This one's a goner," he said. He wrinkled his sunburned nose and tensed his chin into a stoic pout.

I nodded.

"I'll grab the shovel," he said, leaving to retrieve it from his truck.

I dug my fingers into the soil and pulled up a couple of roots, which snapped, confirming what we both already knew despite my unreasonable hope. There was no saving this tree.

"I'm starting to have the same problem over by the Groene Loper," Tjeu yelled over the traffic as he walked back. He was shaking his head, rubbing his palm against his white stubble as he swung the shovel in his other arm. I understood the immense stress he must have felt. The Groene Loper was the largest project he'd ever overseen, the kind of historic undertaking that could not only make or break a career but could forever alter the public's appetite for funding urban forest initiatives.

He drove the shovel into the hard soil with a swift kick of his heel and dug with urgency as if we needed to canvass the whole city in a single day.

I prepared the bag and helped him lift the tree into it.

"Why don't you try sensors?" I asked.

He paused, then laughed. "Yeah, right," he said, gathering his composure. "A sensor couldn't possibly tell me something I can't see in the field."

I tried to bite my tongue. I knew Tjeu's frustration well. I'd had a long streak of terrible luck with the tomatoes I planted in my community garden. I was heartbroken each time I found the plants wilted and dead in the mid-summer. The last time that happened was over a year ago, a few months before an urban farmer introduced me to her soil sensors.

We walked silently for a few blocks until we slowed in front of another parched tree. I couldn't stop the thought from pouring out of me: "Tjeu, your trees are *dying* by the dozens—by the hundreds in some sites. What do you have to lose?"

This time, he just shook his head and dug. He gave me the organic waste bag, and I held it open while he heaved the tree inside, its dusty root ball sticking out of the top.

"Maybe you're right," he said.

I was glad that Tjeu might consider the idea. Still, his initial reaction exposed a challenge I would become familiar with. As I'd discover, it haunts every ecologist who advocates for technological innovations.

My research colleagues and I worked with Tjeu to set up an experiment to reveal how the new tool could solve their problems. There were two sites—one on the Groene Loper and the other on the outskirts of eastern Maastricht. At the Groene Loper, we installed three soil-moisture sensors, and at the second, none. The soil type, drainage conditions, tree species, and weather were the same across both sites. To make things more interesting, we settled on a bet: if I won, he'd have to buy me a dish of my favorite *zuurvlees*, a traditional meat stew considered a delicacy in Maastricht. He'd get a week of my free labor during my next summer break if he was right.

We let the tree managers carry on their duties—planting, watering, and pruning—for the rest of the summer. When the sensors we installed registered soil that was too dry, the tree managers received a notification on their phones, alerting them to water a particular tree or grove. The managers could plan accordingly in these locations, organizing their day around where, when, and how much to water.

As we neared the end of August, I assembled the tree managers to see how they fared. Almost a fifth of the trees without a soil sensor had died. At the Groene Loper, not a single tree was lost. Though I'd learned from my tomatoes to expect a result like this, comparing those that survived in the location without sensors to the trees in the other site revealed a surprise: the ones grown with sensors were vastly larger, some of them by up to three times.

Sitting on a bench halfway up the grove in the Groene Loper that had previously suffered perennial tree deaths, Tjeu and I admired the soft, gentle rustling of the Linden canopy above us. Its leaves were broad, heart-shaped, and smooth, with an intense green hue that shimmered below the blue sky, shielding us from the scorching sun.

"This isn't *I told you so*," I said, unable to ignore Tjeu's awe.

He turned and met my gaze. "I don't know what to say." He stood, then laid down on the shady grass, spreading his arms and closing his eyes.

When he rose a few moments later, he said he felt humbled by the results. "I'm proud . . ." he began, "but disappointed."

I understood what he meant. I felt it, too. We sat in silence, thinking about what all those dying trees meant. The cries for water that went unheard. The thirsty roots that had received too much water in one pour, suffocating them with rot. The people who would have to go without shade. The future generations who would be forced to suffer under increasingly worsening urban heat island effects. Tjeu wondered out loud how many trees had died under his watch. How much money could've been saved by making a modest investment in sensors instead of replacing trees every few years? He spun, following a rumbling plane descending across the neighborhood.

When he turned back to me, Tjeu pursed his lips into a narrow smile, and for the first time, I noticed the dimples in his cheeks.

"Let's install sensors all over Maastricht," he said.

Several years ago—thanks in part to this experiment and the chain reaction that followed—I started exploring how technology could help us better understand the complex relationship between urban living and the nature we rely on. It drove me to research all the other ways urban ecologists used technology to turn our understanding of the living world upside down.

At the core of my mission was an understanding formed from my own experience and my interaction with dozens of neighbors: that the limited nature in our midst was an essential character in our lives. Painters and documentarians never seemed to direct their attention to pocket parks or streams bubbling under railroad tracks the way they sought to capture the beauty of a rugged mountain peak or a group of wild horses galloping across a field. But as every city dweller seemed to understand, the wildlife we interact with daily is a breed of nature that may be even more important.

I have spent the better part of a decade chasing the insights that Tjeu took to heart during the day we spent together in Maastricht, challenging the false dichotomy between nature and technology. Instead of treating the two as separate systems, locked in a grisly competition that ends with a barren city, I've been fighting to show that they can and should complement each other. And even more, if we are to restore the habitats that sustain humanity, we must teach these forces how to be friends.

The division between "natural" and "technological" is nowhere as great as in the urban enclaves where our species predominantly live. For decades, the concept of a "smart city" has dominated the urban planning discourse. While technology has benefits in improving efficiency, safety, and innovation in our neighborhoods, we tend to forget that cities are built on the back of nature, becoming ecosystems that interact with the environment.

To build a truly "smart" city, we must prioritize the health of our environment—and in turn, ourselves. This requires confronting the trends of modern development that have destroyed natural habitats and caused air pollution and rapid deforestation. It means grappling with the effects of climate change and the severe consequences of our nature-starved environments regarding our health and well-being. But first, it demands a fundamental shift in how we view our place in the world—recognizing that we are not separate from nature but a part of it.

As my research revealed, the soil sensors that I showed Tjeu were just one prong of a new industrial revolution remaking the landscape of our cities. Innovators I'd never heard of just ten years ago are leveraging emerging technologies like artificial intelligence, automation, blockchain, and sensors to tackle the problems that routinely destroy our habitats—our homes and the natural spaces around them that make them livable. I decided to devote myself to this emerging science, first by studying it for my doctorate, then by identifying the new field. I call it the Internet of Nature (IoN) because it draws from technological break-throughs to support habitats for humans and all other species to benefit both.

Nature takes on special significance in cities, bringing energy, life, and balance to the environment, providing an essential contrast to built spaces. Like a conductor leading a symphony, nature directs the flow of activity and connection between people, places, and creatures. It provides the oases where we can escape the pressures of work, parents, and kids. Its miraculous principles facilitate the treatment of our drinking water. It filters our air, reduces noise pollution and crime, mitigates flooding, and promotes physical activity. To live in harmony with our environment is fundamental to our survival.

Yet, at alarming rates, urban development continues to gobble up nature in cities, leaving ever-shrinking landscapes for refuge. Concrete walls and new buildings rise where meadows and wetlands once provided shelter for creatures like birds, honeybees, and squirrels. Disappearing parks deprive us of a much-needed connection to the natural world. In this age of towering skylines, the landscapes where more and more of us live are increasingly devoid of life. Uprooted from the environments in which we evolved to thrive, it is no surprise that so many people feel lost in our hometowns—our souls thirsting for the way things used to be. Despite being surrounded by bustling streets, many feel disconnected from the world around them. As I have found, the consequences of this detachment are not just emotional: they cut to the core of our being, forcing us to question our very existence. It turns out that finding our place in the modern world involves reconnecting with our natural environment.

I, like you, had questions. As cities continued to expand, deepening into the earth and reaching the skies, how could we recreate the rich environments in which our ancestors thrived? Could we build a cityscape where manufactured structures and ecosystems made each other stronger? How much time did we have before it would be too late?

Though I would not find easy answers, my research forced me to build from a few unavoidable conclusions. Perhaps most apparent to all of us who try to thrive in our changing environment, there is no doubt that we can no longer ignore the clash between humanity and its wilder elements. Rising temperatures, extreme weather patterns, and their costly impacts on our cities' basic services—infrastructure, housing, livelihoods, and health—are signs that our planet is in peril.

The ensuing chaos is a stark reminder that we aren't in control as much as we once believed. In these moments of desperation, we must find the resolve to push forward.

The question isn't *if* we'll change but *how*. We have the power to determine whether it will be by design or by disaster. Our threatened state demands that we forge a future that honors ourselves—creatures of civilization—and those we cannot tame.

Today, more than 4.4 billion people live in cities—56 percent of the world's population. The United Nations projects that number will rise to 80 percent by 2050. What is less known is where this urban growth is happening. Yes, urban centers are growing, and they will continue to. But many of the 4.4 billion people "living in cities" would consider their home a suburb. And this group grows faster with each passing year. In addition to a great *urbanization,* we are also witnessing a great *suburbanization.* From Chicago to Chennai, India, and beyond, populations are flocking from downtown areas and the countryside to the city's margins, developing metropolises stretching further toward all corners of the globe.

The source of this trend is one to celebrate. History has shown that the world becomes richer as science and technology advance, lifting millions out of poverty and thrusting them into the middle class. As wealth grows, so does the appetite for space. It's a sign of prosperity that our cities have sprawled so far.

But despite its venerable origins, the effects of the trend toward sprawl are devastating.

One day, I saw them myself—in the town where I grew up. Born in the Netherlands, I moved with my family to Canada in the late 1990s: to Waterloo, in Ontario's Southwest. The dot-com bubble was booming, and my father had been given an opportunity to be a part of it. With my mother and baby sister in tow, Canada's tech hub became home.

Officially dubbed the Regional Municipality of Waterloo, it is a metropolitan area containing three cities in the heart of Ontario's greenbelt, the location of the province's most arable land. The population of the region was about 350,000 when we moved there. Now, twenty-five years later, it's nearly doubled.

I grew up beside a small patch of woods, on the other side of which lived one of my best friends. Every morning, we'd coordinate our daily walks to school, communicating on walkie-talkies when we would meet on the "back path," which connected our houses' backyards, separated by some 300 meters of forest. We'd walk the remaining kilometer from the old tree stump to school together. A first-generation immigrant struggling to learn English, I treasured our morning conversations. They were opportunities to try out the new words I'd picked up without risking ridicule.

The truth was that my situation wasn't as unusual as it sometimes felt. The city was home to diverse families from all over the world. It was normal for half my classmates to be born outside of Canada. And as different as we may have seemed, we all shared at least one thing: our archetypal suburban environment—a mosaic of cul-de-sacs with no sidewalks, woven between large tracts of forests and meadows that developers had had the foresight to preserve. We yelled to our parents over the hum of lawnmowers from atop the same species of trees, refusing to come in for dinner. We propped up sticks against the trunks, making small forts that dotted the woods. In the winter, as soon as the smallest flurry edged into the forecast, we prayed for snow days. We would borrow the lids of trash cans to sled down the snow-packed hills. There were no wailing police sirens or drunken fights outside our windows. Nothing ever happened there. And that's precisely what made it so great.

Then, I saw it all change. The house I grew up in used to be on the city's outskirts. Some 500 meters north of my house were woods, cornfields, and farms as far as the eye could see. But in my nearly two decades there, development cannibalized the landscape. Where fields of prairie flowers and grasses once blossomed in June now stood soggy plots punctuated with identical beige homes. In the place of ancient trees had sprung saplings—meek attempts to replace what was lost under suburbanization's swath of destruction. Just as my childhood home had helped to engulf another landscape in urbanity's tide, it was now locked within the city.

Walking the family dog through these new neighborhoods when I visited my parents for the holidays in the years after I moved away, I questioned everything about the environment I'd once called home. Why do we honor our trees, lakes, and flowers in the names of our towns, neighborhoods, and streets but struggle to conserve those very organisms? Why do we have to choose between nature and development?

A growing movement of urban planners and engineers recognizes that this age-old dichotomy is misguided. As hunters and gatherers that grew up in the forests and savannas, we cannot thrive removed from those spaces. Nor can we completely immunize our surroundings to the dangers of the wild and the earth's vicissitudes. As we come to terms with sprawling urban landscapes and ready ourselves for the effects of climate change, nature within city limits is an often-overlooked asset in our arsenal.

From forests to parks, stormwater infrastructure to waterways that nourish us both physically and emotionally, these rich sources of life can be leveraged as powerful tools to strengthen our habitat's resilience. Early urbanites have known for centuries that nature keeps us grounded as cities evolve. Indeed, many of the parks we enjoy today—London's Hyde Park, New York's Central Park, Boston Common—are the same parks where our ancestors sought refuge and respite from the harsh realities of city life. As early as the eighteenth century, the phrase "Lungs of London" was coined to describe the city's trio of Royal Parks—St. James's,

Green, and Hyde—which offered an environment that would mitigate the shortcomings of the smog-ridden Industrial Revolution and subsequent urban sprawl. Against all odds, nature thrived in these human-dominated environments, and in turn, cities became tolerable habitats for both people and wildlife.

In broad strokes, what some intrinsically understood in the eighteenth century is still true today: where nature thrives, people thrive. Only now do we have the evidence to back up the observation. Yet, as urban development creeps worldwide, nature must fight—and often die trying—for its rightful place in the city. Suffocated by asphalt, cut by utility lines, and blinded by the lights of our cities, nature can no longer support the burden we put on it. More than just a place to hide out from dirty air, we now lean on our ecosystem to cool our streets, absorb carbon emissions, and buffer neighborhoods from rising seas, flash floods, and wildfires. And there is no denying that it is falling short.

Global population growth and the growing middle class is one reason nature cannot keep up with what we ask. It fulfills a laundry list of duties when cities and towns burst at the seams. Only 6 percent of the world lived in cities when the first urban parks were built. In North America, that number is now nearly 85 percent. For the urban capacity to keep up with population growth, nearly 80,000 km² of land—roughly the size of Austria—is urbanized each year.

Nature's inability to sustain population growth has consequences. Every week, three million people across the globe relocate to cities, often seeking prosperity only to find themselves threatened by challenges their new homes weren't built for. Once unprecedented heat waves now routinely cripple aging infrastructure designed for a different climate. Sunny day floods teem out of coastal sewer systems constructed in an era when storms weren't as big and prairies used to dot the landscape, diverting rainfall. Thronged natural spaces have failed to reconnect us with the wildlife surrounding humans for hundreds of thousands of years, threatening our well-being. Each new reality jeopardizes our mental and physical health.

While some of these challenges only became apparent with the emergence of climate change, the germs that gave rise to them have been with us for decades, ignored by planners and municipal leaders who subscribed to an ill-fated ambition to sever our connection with the ecosystems around us. If we could manipulate the wetlands with concrete, the thinking went, create wider roads to ease a clogged commute, and insulate ourselves from the weather using air-conditioning, we could divorce ourselves from the ferocity of our environment. Nothing would be beyond our control. But as nature begins to overpower the antiquated systems that came out of this mindset, it forces us to confront the disastrous costs of the fallacy. Every day in cities worldwide, millions now swelter under extreme heat events, develop diseases such as asthma due to poor air quality, and buckle under new mental stresses imposed by urban life.

The conundrum that compelled my research into the IoN is existential: How do we restore the balance that once characterized human environments without losing the progress we have made? Rightly, many celebrate the urban shift that the Industrial Revolution unleashed. The technological explosion that began with factories springing up along rivers lifted billions out of poverty. It fanned across the globe, drawing people from the countryside into highly productive hubs. The vast social networks it unlocked accelerated innovation. And the smaller homes that urban density facilitated now enable us to reduce each person's carbon footprint. But with so much focus on the miracles of the modern city, our leaders have ignored the bills coming due. Now, at the dawning of the urban century, a growing pile of stumbling blocks endangers the whole experiment.

As we enter a new era characterized by ecosystem restoration and building nature within city walls, it may be tempting to look back at what we've lost. The majestic species that succumbed to their changing environment and the ecosystems overrun, the days when human meddling left nature untouched. But the journey to build sustainable, healthful habitats isn't about "going back to nature" but about moving forward into nature. To succeed, future development shouldn't replicate old environments for nostalgia; rather, relying on the lessons of modern life, it should import into our habitats the elements we need to thrive.

It won't be easy. The lack of sufficient research into the impact of overlooked green spaces presents one major hurdle. Despite the invaluable benefits of urban nature—both to people and our planet—its economic worth often goes unrecognized. Local governments shoulder much of the financial burden for these benefits, while an array of beneficiaries—like property owners, businesses, and insurers—get a free ride to enjoy the rewards. It should therefore be no surprise that those with control over the purse strings often struggle to make a compelling financial case for investments that will save lives. To cultivate greater investment in cities' ecosystems requires first helping nature to speak our language. We hope to turn the corner only when cities can properly quantify the effects of supporting ecology.

Physically, there are other challenges. Urban environments can be downright hostile, even for stalwart organisms. Pollutants, soaring temperatures, and lack of adequate space above or below the ground all pose obstacles to proper growth—making plants liable to infections and afflictions. To overcome these challenges and foster a vibrant urban landscape, city-dwellers or their governments must address the often-unclear ownership responsibilities for vegetation, organizing groups that care for public nature. Solutions begin at the local, neighborhood level. While uncovering why some neighborhoods had made astounding progress while others suffered, I discovered a hidden truth embedded in their success: technology. The same technology we once shunned and feared as the "enemy of nature" has now emerged as its strongest ally.

In our quest for answers, we need to explore every possible solution—from the transformative powers of technology to the timeless wisdom found in natural systems. For all the damage that human exploitation wrought, the technologies it built are now our greatest advantage for reclaiming the ecological networks that once supported us. We cannot, for instance, remove the sewer pipes webbed beneath our cities and restore the marshland beneath the street. But we can map the flow of aquifers to make our drainpipes effective once again. As the Maastricht experiment taught me, tomorrow's cities must rely on emerging technologies—and repurposed older ones—to build durable ecosystems.

Finding harmony between nature and technology lies at the heart of the IoN. As cities grow and modernize, so must the solutions for protecting our natural resources. In this book, I learn from the visionaries around the world who are bringing these old enemies together to build a habitat where humans flourish.

The work of these inventors and entrepreneurs also shows how to build the necessary support to unite nature and technology. Before city governments approve the extensive investments necessary, researchers must prove the value of the right technologies. For ecologists who've witnessed decades of human engineering that attacked nature, that's a tall order. Many nature lovers have watched industrialization trample our ecosystems, developing an understandable skepticism of the latest innovations. The visionaries that I sought out learned from the dangerous oversights of modern development to illuminate ways in which technology, when subordinated to the miracles of Earth and used to support them, can unlock the value of urban nature. Their stories offer vital insights into building resilient, vibrant neighborhoods where the wellness of the residents and the planet go hand in hand.

Ultimately, the question future generations are facing is simple: Confronted by the twin crises of urban growth and climate change, how can we create communities of abundant nature, utilitarian wildlife, and everyday convenience where verdant flora and bustling fauna seamlessly support the comforts of modern life?

As it turns out, the answer starts with recognizing the splendor of urban nature's most conspicuous element—its city-dwelling trees—which has been too often ignored. Only then may we appreciate and value these silent sentinels for all they provide us: shade and shelter from the sun's harsh glare, a connection to our past, and a buffer against the storms and smog of our future. So that's where this book begins—revealing the perils of urban deforestation, the misguided replanting efforts, and the opportunities for a new approach.

The urban forest measures the weight of careless destruction and charts the journey to renewal. Beneath every cut tree is an armada of missed potentials—root systems decimated by their plight and saplings that never grew. By understanding which technologies give us a puncher's chance at deciphering nature's messages, we can save this valuable resource before it is too late.

To hear the cry of a thirsty tree or the wail of a river preparing to breach its levee—to truly understand their meaning—we must first learn how to listen.

CHAPTER 2

Saving Trees
on the Paper Trail

Technology won't replace arborists but arborists who use
technology will probably replace arborists who do not.
—Josh Behounek

"BOSTON PARKS DEPARTMENT, HOW can I help you?"

I was standing at the tree warden's office door, watching Superintendent Greg Mosman twirl the scrunched cord of the old office phone he was using, leaning back in his creaky swivel chair. He gestured to me to give him a second. I heard a high-pitched, almost whiny voice on the other end of the line.

I'd walked into several of the wrong buildings trying to find Greg's office. Though it was only steps away from Fenway Park, the legendary home of the Boston Red Sox, it was tucked into the corner of a small green space, the windows covered in black steel grates. When I finally noticed the structure, only a small sign that read "Boston Forestry Division" told me I was in the right place.

The inside of the building was dark and musty, with fluorescent tube lights jutting from the low ceilings close enough to make you want to duck when you walked under them. It reminded me of the portable classroom I spent fourth grade in. That was supposed to be temporary like I imagine this shell of a building must have been. In the Canadian winter, my classmates and I sometimes wore coats to withstand the room's spectacular draft.

The town had built those classrooms to keep up with the growing number of children in the neighborhood—what leaders at the time believed would be a brief growth spurt, but the trend never seemed to stop. When I walked by my old primary school last Christmas when I was visiting my parents, the portables stood exactly where they'd been twenty years ago, housing a new generation of tweens.

A lack of municipal planning wasn't surprising in a once small town overtaken by a tsunami of sprawl. But the fact that those responsible for creating and maintaining more than 160,000 trees and 2,300 acres of parkland in Boston operated out of this space spoke to something else: a failure to understand the value of nature as an essential social good.

Greg's chair cracked like an arthritic knee as he leaned forward. "Yes, no problem," he told the caller. "Can you tell me which tree it is?" He wheeled over to his desk and pulled up Google Maps.

"Let's try the address," he suggested. "Riiiight, okay, so, errrr, between seven and nine on Marlborough Street. "Hmmm, yeah, I think this is an old picture," he said, peering over his glasses at the Google Street View image. "Okay, I think I know the tree—I'll try to get one of my guys out there by the end of the week to look at it." After a pause, he insisted that was the earliest option. "We're completely . . ." he said, trailing off.

He sighs and looks over to me, forcing a grin. "The guy hung up on me!" he said. "Anyway, sorry to keep you waiting."

I introduced myself, then a Ph.D. student looking to take soil samples of mature red maples. After some nervous fumbling, I got to my question: "Can I use your tree inventory to weed out the locations of red maples?"

He cast his eyes down as if studying his shoes, muddy from so many strolls into the parks. "Unfortunately," he said, clearing his throat, "we don't have a functioning tree inventory, never have. But I can note trees that fit your criteria as I drive around the city if it would be helpful."

I was incredulous. "You can identify whether or not it's a red maple from your *car*?" I immediately regretted asking such a dumb question.

Greg chuckled. Luckily, he was not offended. "After working as the city's chief arborist for so many years, you more or less know where each tree is." For a city of nearly 200,000 public trees, the claim seemed outlandish. My eyes turned to the scribbles on his hand, the hundreds of Post-it notes on the wall behind him, and the piles of paper on his desk.

"I love my work," Greg continued, following my gaze. "And I've never met an arborist I didn't like. We're die-hard tree lovers and know the urban forest like the back of our hands. But as you can see, that's sometimes a problem. Many of us are keeping track of their trees with Post-its." He pointed to the pen-etched scribbles on his hand. "Or worse."

A couple of days later, Greg emailed me a list of twenty red maple locations organized by address. If you ever come up with a way to better manage trees, he wrote in a postscript, call me.

The experience got me wondering about the prevalence of tree inventories. If a city like Boston, with a park system designed by the celebrated landscape architect Frederick Law Olmstead, didn't have a tree inventory. How many other cities worldwide were trying to manage their ecosystems in a data vacuum?

As I would come to appreciate the astonishing talents of arborists like Greg in the coming years, I also came to see how the absence of municipal investment crippled their work. In lacking a tree inventory, Boston, I found out, was not the exception but the norm in city governments. And without the data to show the impact of the urban forest—the trees, shrubs, and associated vegetation across a city—officials and citizens structurally undervalued it, often dooming trees through poor planting and caretaking decisions. The consequence is a city either devoid of a tree canopy or dominated by an urban forest monoculture that is vulnerable to rapid depletion.

Over the past few years, mass mortality events like Dutch elm disease, chestnut blight fungus, emerald ash borer, and oak beetles have taught many of us about the dangers of these monocultures: they allow the species- or genus-specific pests and diseases that global trade facilitates to destroy an entire city's forest. The mass mortality events resulting from our homogeneous urban forests are predicted to wipe out 1.4 million urban trees by 2050, requiring a staggering cost of over $900 million for replacements. Hot spots that are particularly vulnerable to this tree mortality include Milwaukee, Chicago, and New York City.

We are only beginning to feel the effects of these losses. As their jarring absence reveals in cities that have failed to manage their forests, the benefits of having trees in cities extend far beyond mere aesthetics.

You might be most familiar with the ability of trees to absorb pollutants and produce oxygen, thus improving city air quality. Trees decrease the risk of respiratory diseases such as asthma and chronic obstructive pulmonary disease (COPD) by reducing nitrogen dioxide, ozone, and fine particulate matter. A study by the U.S. Forest Service found that trees removed approximately 17.4 million tons of air pollutants in the United States each year, providing an estimated $6.8 billion in economic benefits to Americans. Additionally, trees provide vital shade during the summer, reducing the need for energy-intensive air-conditioning. A Toronto and Region Conservation Authority study found that a single mature tree offers the same or better cooling benefits as five air-conditioning units running for twenty hours daily. Trees also release water vapor through a process called transpiration, which helps to keep the air around them cool in the same way that sweat chills the surface of our skin.

Lush greenery is also vital in protecting cities from the growing perils of relentless rain. Trees, particularly their leaf canopies and roots, serve as natural sponges, absorbing substantial amounts of rainwater and curbing floods. According to the U.S. Department of Agriculture, a single mature tree can absorb up to 1,000 gallons of water annually.

By absorbing pollutants, releasing oxygen, and regulating temperature, trees make urban areas habitable for other organisms that support human life. They also provide a haven for wildlife that cannot survive only on the concrete and asphalt surfaces that otherwise define the city, contributing to ecological diversity. Trees offer shelter, food, and breeding grounds for these species, often in synergistic relationships that benefit both the tree and the resident species. Without trees, urban areas would lack the rich array of life that, I would learn, is necessary for the health of city residents.

In part because of the biodiversity they facilitate as well as other benefits, such as dampening noise pollution, the oasis that urban forests provide have been linked to improved health outcomes. Researchers have noted their calming effect on people, reducing stress and anxiety. Exposure to green spaces has been shown to improve cognition, increase social connections, lower blood pressure, reduce symptoms of depression, and improve immune function, decreasing the risk of chronic diseases and disorders.

The incredible influence of trees on our quality of life may not be surprising when we consider the history of humanity. Throughout our evolution, forests have played an essential role in shaping our existence, providing shelter, the basis of our fruit-dominated diet, and materials for making the tools that set us apart from the rest of the animal kingdom. No wonder many people feel a visceral connection to these pillars of nature.

Why, then, have we struggled so mightily to preserve the neighborhood trees that we rely on? Seeing Boston's approach through Greg's eyes, I discovered that our approach to managing urban forests played a major role.

Tree inventories, which compile data on the number, species, age, location, size, and health of the trees in an area, are the most important tool for caring for urban forests. They may seem like a luxury to city governments that are focused instead on water treatment, building inspections, sewage, and crime. But they're the foundation on which resilient urban forests are built, and a burgeoning field of research shows urban forests may be just as important as a city's other essential functions for saving lives. They can help forest managers improve our cities by highlighting the areas that could benefit most from the shade, water absorption, and purified air of new trees. They can also protect our towns from decline, providing the granular data that forest managers need to protect these critical resources from negative impacts such as felling, pests, and disease.

Without adequate inventories, we cannot address the risks posed by dead limbs, which can hurt residents traveling beneath them or living beside them; we leave our cities more exposed to disastrous flash floods; and we lose an essential weapon against the threat of the urban heat island effect, exposing residents to killer heat waves. Then, in the wake of natural disasters, the lack of an inventory thwarts recovery efforts. Deerfield Beach estimated it lost thousands of trees when Hurricane Wilma violently struck South Florida in 2005. But since it didn't have a tree database, the city struggled to qualify for reimbursement from the Federal Emergency Management Agency (FEMA).

Dr. Andrew Koeser, one of the world's foremost tree inventory researchers, recently surveyed to understand these risks. He found that U.S. cities without functional tree inventories lag their counterparts on metrics such as removing hazardous branches, prioritizing maintenance, identifying storm-damaged trees, charting species diversity, and estimating ecosystem services, the free benefits that nature provides us.

Since I met Greg five years ago and began to make sense of the plight of the urban forest, one important thing has changed. After dealing with the cost of dead branches tangled up in phone lines, disposing of dozens of forgotten sidewalk saplings, which died from a lack of water, and cleaning up the destruction left by a forest of genetically identical trees overcome with a disease, many cities now understand the benefits that a tree inventory could provide them. What holds them back today is not desire but a long-held belief that keeping such a database will require a staff of hundreds. Officials imagine a swarm of arborists circulating every city street and park with pen and paper, recording what they find. In fact, this is how 60 percent of respondents to Andrew's survey collected data, a process so slow as to render the inventory permanently out of date. Although the practice of manual tree management persists, new technologies present a much cheaper, more effective alternative.

Singapore is one of the world's greenest cities. With seven million trees and one million more expected to be planted by 2030, Singapore's inventory is a "digital twin"—developed by capturing measurements using light from a laser directed around each block and then applying artificial intelligence to geolocate each plant. Arborists at Singapore's National Parks Board (NParks) can track their trees through an app and conduct several digital inspections, like examining overall tree stability under fierce tropical storms. It has a network of physical tilt sensors, which measure sudden movement and indicate risk, remote sensing to determine healthy chlorophyll levels, and street-level cameras for remote visual checks—all of which play a role.

In 2000, NParks reported about 3,100 "tree incidents"—where a branch falls, or a tree trunk snaps or is uprooted. Most notable in the years since was a 270-year-old tembusu tree, which fell during a concert in the Singapore Botanic Gardens, crushing a family. Concertgoers rushed to lift the massive tree—saving the father and one-year-old twins, who suffered injuries. But it was too late for the children's thirty-eight-year-old mother, who was killed. The tragedy showed the community needed a better way to manage its forest.

Over the next years, the city focused on continuously improving its tree database and tree care, eventually leading to the inventory model and its companion app. By 2020, NParks reported a nearly 90 percent decrease in annual tree incidents. By identifying potential safety hazards and assessing each tree's overall health and maintenance needs, NParks proved it could address issues before they became dangerous to people and property. It restored confidence in the safety of its trees that the Botanic Gardens incident had shaken, building public support for growing the urban forest after years in which some residents had begun to question the value of trees. In doing so, it provided a model for cities worldwide, which struggle to field all the complaints from residents worried that an unkempt tree might harm them or their homes.

The movement that Singapore championed to value and manage the urban canopy properly had begun years before in New York City. The most densely populated city in the United States, its soil stuffed with thousands of cubic feet of "utility spaghetti," New York started its tree census practice in the 1990s. Still, the project had taken on much greater significance after the city began to appreciate its remarkable forest anew following the World Trade Center attacks.

The story started a few years before that, with a public effort to count every tree across the city. The city's parks commissioner had delegated the action to Fiona Watt, a recent forestry graduate who had only started working for the parks department five days before. Though she felt completely unprepared to lead what may have been the world's first urban tree census in a major city, Fiona wanted more than anything to prove that it was possible.

Years before digital mapping systems and sensors could automate the process, Fiona knew she would need an army to canvass the city. So she and the longtime parks commissioner, Henry J. Stern, invited the public to help. At the press conference, Commissioner Stern asked anyone willing to help count trees to call a new hotline: 1-800-360-TREE.

When people called, Fiona was on the other end of the line, madly scribbling down their names and phone numbers. Within days she'd amassed an infantry of two hundred volunteers. It launched a new era for NYC Parks, in which citizens were grossly more engaged in the trees they shared their streets with.

Fiona used her forestry background to determine the variables to measure in the field. It started simply with the location, species, size, and condition. Everything was done by hand. Without the online maps we rely on today, Fiona had to use road atlases, which she annotated for her new community workers, laminating the paper to protect it from the elements. Each volunteer received a map highlighting the polygon they were partly responsible for and a stack of data-collection sheets on a clipboard. Later, the city contracted a company to keypunch all recorded information into a rudimentary database. It was the first technological leap for the newly dubbed "TreesCount!" campaign.

The first tree census revealed surprising truths. For starters, 10,000 dead or dying trees had gone unnoticed. The observation sparked new programs devoted to pruning trees and replacing the dead with stronger saplings. Perhaps even more surprising, the census helped to explain why so many New Yorkers felt deeply connected to the urban forest: it was far more widespread than anyone realized. In a city better known for its glass and steel towers, trees outnumbered the skyscrapers more than 500 to 1.

In the weeks following the September 11 terrorist attacks, the importance of New York City's parks only became clearer. With the city mourning the 2,753 fallen, parks emerged as a place of healing. In a letter to the public just two weeks after the attacks, Commissioner Stern wrote about people lighting candles, writing messages, leaving flowers and flags, and consoling one another in their neighborhood parks. "Our city has sustained a tremendous loss," he said, "but with your help, parks can play an important role in the psychological recovery."

Fear, economic downturn, and respiratory conditions linked to exposure to the World Trade Center collapse—which would eventually claim thousands more lives—had residents fleeing the city. When Michael Bloomberg became mayor three months later, smoke was still rising from downtown. Many people wondered whether Lower Manhattan would ever come back.

But as residents continued to filter out and the city mourned, its trees continued to purify the air. Some reminded the public of the resilience we all share within us. When the World Trade Center site was developed as a memorial park where those who had lost loved ones could reflect on the tragedy, its designers transplanted a Callery pear tree at its center. The tree had been discovered among the rubble, nearly incinerated but still showing signs of life. Workers nursed it back to health for a decade, sprouting new branches and flowering in spring; it grew from eight to over thirty feet tall. With limbs extending from gnarled stumps, the "Survivor Tree" experienced what so many New Yorkers did that day. Badly burned in the rubble, it remarkably recovered, sprouting new growth the next season. As one survivor of the attacks reflected, "[it] reminds us all of the capacity of the human spirit to persevere."

When Fiona returned from her third maternity leave to tackle the city's next tree census in 2005, it was clear the stakes had grown. Her colleagues at the Forest Service's New York City Field Station had begun to document how New Yorkers created "living memorials" or green spaces dedicated to memorializing those lost in the 9/11 attacks. Her neighbors across the city were more tuned into their reliance on nature than ever, with new organizations devoted to nature popping up yearly.

So she tapped into that enthusiasm, using promotional campaigns to help quintuple the tree-counting infantry. This time, she also hoped that better technology could make the volunteers more efficient. Instead of just pen and paper, she distributed handhelds with a rudimentary web app to her more than one thousand volunteers, who could now immediately input data into the system. At the end of 2006, volunteers surveyed over 590,000 street trees in excruciating detail. Next, Fiona called for the help of the one person she knew could motivate the city to use this data revolution.

Dr. Greg McPherson, Fiona's onetime professor, was a uniquely urban-focused researcher with the U.S. Forest Service. Calling himself a "green accountant," he researched new methods and tools for quantifying the value of nature's benefits from city trees, creating a model with which leaders could battle for the budgets they needed to maintain their trees and expand urban forests. Dr. McPherson called the model the Street Tree Resource Assessment Tool for Urban Forest Managers (STRATUM), which was later adapted to a state-of-the-art, peer-reviewed software suite called i-Tree Streets and, more recently, i-Tree Eco. To this day, Dr. McPherson's model remains the gold standard for thousands of communities, nonprofits, consultants, volunteers, and students worldwide to calculate the benefits of trees.

The model hadn't yet launched when Fiona called him. But having heard about it, Fiona knew it could give her exactly the data that New York needed for city leaders to understand the role of trees in people's lives. Dr. McPherson told her it wasn't ready. It was early days, and he hadn't completed the reference studies for New York—or any East Coast cities, for that matter. Since the model was based on West Coast tree species, growth patterns, and climate conditions, it would be useless, he said.

Not one to take no for an answer, Fiona secured a grant from the state to adapt and apply the model to New York City. With Dr. McPherson's engineering, it worked. Within a few months, Fiona got the results that would forever change the face of the city's forest: for every dollar spent on tree planting and care, trees provide $5.60 in benefits, capturing carbon dioxide in their tissue, reducing energy used by buildings, decreasing air pollutants that can trigger asthma and other respiratory illnesses, and acting as natural stormwater retention devices, with the combination of these benefits increasing property values. For the first time, the NYC Parks Department could translate the value of the city's trees into dollars and cents.

Fiona immediately took the results to her boss, Adrian Benepe, the parks commissioner who followed Stern, who then presented it to Mayor Bloomberg. As a former investment banker turned entrepreneur, the mayor had stacked his administration with ex-executives and bankers, people who appreciated return on investment (ROI) like no other administration before. As Adrian presented the data, all the administration heard was a 560 percent ROI. It paved the way for the biggest investment in street trees the city had ever seen.

It took the form of a collaboration with Bette Midler, who'd recently founded a nonprofit dedicated to cleaning and restoring green spaces in the five boroughs called the New York Restoration Project. Inspired by tree-planting campaigns she'd seen in other cities—from Los Angeles to Denver to Boston—Bette, who had moved to New York City, felt the city needed to do something even more ambitious. While walking in the park with the mayor, for whom the ROI data was undoubtedly top of mind, she told him about her vision for her organization. Together they conceived the MillionTreesNYC initiative.

Fiona told me about the program while we stood in the shadow of a thirty-some-foot-tall Carolina silverbell in the Morrisania neighborhood in the South Bronx. It was a warm morning in September, and we stared up at a cardinal gliding down to a nest of squawking chicks. Bette had picked this location for the first tree that the initiative planted. One of the areas with the city's highest asthma rates, the intersection at East 166th Street and Washington Avenue was otherwise starved of vegetation.

The MillionTreesNYC initiative was supposed to take ten years but fueled by a new purpose among city workers and tens of thousands of volunteers, it finished two years ahead of schedule. Not even two fierce storms, Tropical Storm Irene and then Hurricane Sandy, which created months of cleanup work for tree crews, could slow the momentum of MillionTreesNYC.

Most importantly, after years of disinvestment in the nature of less affluent areas, the initiative helped to restore equal access to trees. Of the million new trees, the Bronx received 280,000; Brooklyn, 185,000; Manhattan, 75,000; Queens, 285,000; and Staten Island, 175,000.

By 2015, when Fiona oversaw the third iteration of the TreesCount! campaign, the initiative's success was readily apparent. Notwithstanding the effects of Hurricane Sandy three years prior—the most devastating weather event in the city's recorded history—tree numbers had surged 12.5 percent from 2005 and 33.6 percent from 1995.

Fiona had been able to measure this success in part because she continued to innovate the census. In 2015, she implemented geographic information systems (GIS), a mapping technology specialized for storing metadata such as tree size and shape. With GIS, the census could collect nuanced data showing the campaign's massive investment benefits.

As Jackie Lu, the inaugural director of data analytics at NYC Parks at the time, explained, each census needed to build from the last. The first had supported implementing a block-by-block tree maintenance program, allowing the city to prune trees on rotation for the first time, intervening before dead branches became dangerous. The second census enabled the calculation of the ROI of the city's street trees, making the business case that launched MillionTreesNYC. It forced city leaders to take the Parks Department seriously for the first time.

The third census, by contrast, was a test case, evaluating if the benefits of new trees promised by the 2005 census could actually be realized. At the heart of the latest initiative was a new technological leap—a tablet- and smartphone-based tallying system that would make the job of "volun-*treers*" faster, less labor-intensive, and more accurate. In the past two censuses, the mapping was based on addresses, and as a result, the locations where canvassers pegged trees were often imprecise. Since nearby trees can often look alike, the data confused the foresters who would later use it.

Jackie and Fiona's challenge was designing a recording method simple enough for nontechnical volunteers yet accurate enough that foresters could finally trust it. A local nonprofit called TreeKIT stepped in to help. It designed an app combining century-old technology—a measurement wheel similar to the one land surveyors push around on an axle—with modern-day GPS capability. Using the GPS's orientation and human measurements, the wheels could record each tree's exact location relative to the surrounding intersections and buildings. Jackie's quality control confirmed that volunteers with the device mapped trees almost as accurately as her staff.

Soon, NYC Parks became the nation's largest purchaser of surveyors' wheels. The effort took two years to complete.

Even then, it wasn't done in Jackie's eyes. To keep volunteers engaged for years to come, she wanted to ensure she let them in on exactly how—and *why*—the data they were collecting was used. So after operationalizing the data and bringing it into the software NYC Parks uses to manage the urban forest daily, she made a public-facing version. The NYC Street Tree Map is now the model for cities with similar ambitions.

Continuously updated by the foresters who use the backend, it offers an interactive tool designed to be a central hub for everyone interested in their neighborhood trees. You can identify the ecological benefits of your street tree, contribute data, track the tree's care, organize your own tree care group, and—of course—submit service requests to the city.

As Fiona and I leave the shade of the Carolina silverbell that started it all, we wade through a sea of cars waiting at a stoplight and further into the Morrisania neighborhood. I immediately feel the sun's heat, a pulse on my forehead. Fiona glances back when she makes it to the sidewalk as if to make sure I'm still there, and she squints to see me through the glare. Her dirty-blond hair is pulled back in a tight ponytail, a few abandoned pieces hanging down her cheeks.

We both turn around to face the tree, following the line of the thin trunk reaching toward the top of the brownstones. Taking in the sight of this oasis in a busy intersection, I wonder whether other cities could replicate the success of the TreesCount! campaigns. For the first time, there is finally money to do so. As of August 2022, the U.S. federal government is making a whopping $360 billion investment to address climate change. Forestry projects have been allocated a $5 billion slice of the pie, with $1.5 billion set aside for the U.S. Forest Service Urban and Community Forestry program, which distributes funding to local governments and community organizations. To capture this funding, cities need accurate tree inventories, so they are motivated to get their hands on accurate, cost-efficient data. But will they have the community engagement and department-led expertise to build a database from scratch—and more quickly than New York City ever did? After all, with funding starting to be distributed, they don't have the years that each TreesCount! census did to catch up.

"So what's next for the 2025 census," I ask, trying to distract myself from the hopelessness that I can feel overcoming me.

She tells me she's excited for a new technology she heard about at a recent conference. "LiDAR," she says, "but, you know, on the ground . . ." She looks at me for help.

"Terrestrial LiDAR?" I offer.

"Yes, that's it. I've heard cool things. It could help cities get incredibly granular tree inventories more quickly." She was describing the technology that would eventually facilitate Singapore's campaign.

The first time I saw LiDAR up close was at home in Amsterdam two years before. I was walking down an elm-lined canal with Dirk van Riel, a former tree inspector who spent fifteen years going neighborhood by neighborhood, block by block, to inventory trees. We were reminiscing about our respective experiences conducting tree surveys.

For the past twenty-five years, annual tree inspections have been mandated by Dutch law, making tree inspection time a frenetic and sometimes dreaded season for tree-care professionals nationwide. While Dirk had been leveraging the benefits of satellite imagery, they could only take him so far; satellite imagery couldn't tell him the measurements he needed for a full tree inspection, he said, pointing to the elm closest to us.

He sputters in staccato, like a motorcycle revving. "I need to know things like trunk diameter, condition, crown volume, and height," Dirk adds. So he did the only thing he thought to do: quit his job as an inspector and decided to devote himself to coming up with the tool he desperately needed—a way to automatically detect, measure, and maybe even inspect trees using sensors.

He found that LiDAR, a light detection and ranging sensor already on the market for other uses, could detect the raw data he needed. The technology uses laser beams to measure distances, sending out millions of lasers and then measuring the signal that comes back. Dirk had seen the data used to generate 3D images or maps of an area, providing detailed surveys of terrain and land-use information. He'd heard of others applying LiDAR for self-driving cars, autonomous drones, and robotics applications, where it helps identify potential hazards like sinkholes or avalanche zones. He decided to find a way to adapt it to measuring trees.

Dirk experimented with mounting LiDAR devices on cars and backpacks to map and decipher surroundings. Next, he developed a deep-learning algorithm to filter out everything except trees.

He claimed that his technology, TreeTracker, could use the measurements taken by a car driving down a city street to create a data set of the location, including the trunk diameter, height, and crown volume of each tree the car passed. It seemed outlandish for many who had spent careers calculating these dimensions by hand. Dirk struggled to recruit clients, worrying he'd made the biggest mistake in his life.

Then, a few months later, he started to reach out to cities one by one to show them how they could better manage their trees with a fraction of the staff. Soon, dozens had signed up for TreeTracker.

One municipality took it a step further, applying Dirk's technology to take a detailed census and leveraging the data to determine how best to grow its urban forest. Woerden, a small city of 50,000 south of Amsterdam, had started by mapping its 30,000 trees with TreeTracker in just four days, a process that previously took two years. Then, with an abundance of real-time data at their fingertips, Woerden's planners realized that they could finally compute the answers to the very questions citizens had been asking them—about how much carbon the city's trees sequestered, how much air they filtered, and how much stormwater they captured. Combining the measurements from the LiDAR system with Dr. McPherson's i-Tree Eco model, they calculated each of these tree benefits, which they would proceed to track over time.

Combining LiDAR and i-Tree into a system has inspired a new, more efficient approach to growing nature, quickly spreading beyond early adopters like Woerden, Amsterdam, and Singapore. No longer do these cities measure tree-planting success by the number of trunks it gained or lost—a crude approximation for the health of the urban forest. Instead of looking at such inputs, the cities now focus on optimizing the outputs, which are a function of factors such as total crown volume. An evolution that recognizes that trees are not discrete, interchangeable pieces of scenery but living, growing organisms that deliver specific eco-system services.

As revolutionary as LiDAR is to mapping and maintaining the public urban forest, substantial hurdles remain when dealing with the private sector. Fiona and I walk down 166th Street toward her favorite pizza slice in the neighborhood. We pass a construction site and then a beige art deco tower, its tall, narrow windows surrounded in a layered zigzag of tiles. This way, she says, pointing around the corner.

Walking past the other side of the tower, we're startled by the shriek of a chainsaw. Instinctively, we jog toward the building's edge and peer through the chain-linked fence into the small yard behind it. A small crew is taking down an adolescent tree.

Fiona shakes her head. "This is a huge oversight of our program," she says. "The street trees we focus on are just a fraction of the urban forest."

A 2021 report conducted by the Nature Conservancy points to the scale of this challenge, assessing the urban forest across public and private land. It found more than 35 percent of the city's trees stood on private property.

As Fiona explains, these trees face all the threats of street trees—invasive pests, pathogens, a changing climate—but none of the public protections, including laws prohibiting cutting them down. Having learned the city's history of forest protection, I'm surprised. Parks Commissioner Stern, often credited with coining the term "arbori-cide," had persuaded the city council to pass a public tree ordinance years ago to combat the "murder of trees," threatening fines of up to $15,000 and even jail time. Yet nothing can be said or done to protect private trees from the growing pressure to develop the city's last natural lands. As the Nature Conservancy's 2021 report noted, canopy in all public land uses increased across the city, while private property saw mostly decreases.

Fiona pushes herself away from the fence and starts to lead us again down the block. "It makes me feel helpless," she says, "and frustrated."

It's an emotion felt across the world. As Fiona acknowledges, many other cities have it worse than New York. Estimates indicate that anywhere from 50 to 70 percent of the world's urban forest is on private lands. Meaning more than half is typically out of the municipality's purview.

To get a glimpse of how a city facing the worst effects of private tree destruction was fighting back, I'd need to travel across the country to Santa Monica, California.

Matthew Wells, the forester I wanted to meet there, had made the same trip from New York nine years before me. A tree-care manager at NYC Parks, he understood the scourge of private destruction firsthand, having worked for many years at tree-care companies, where he was often tasked with chopping down the gentle giants he'd learned to love as a horticulture student. When he heard about an opening in Santa Monica, where he could play a major role overseeing the city's urban forest program, he didn't hesitate to take the leap.

Years later, now managing a new private tree-monitoring program, Matt feels utterly at home on the Santa Monica streets. A stubborn individualist, he had found his people.

He certainly looked comfortable in his skin when I first spotted him across the street from the park where we'd agreed to meet. He's standing on a bus stop bench, screening out the sun with a hand on his forehead, peering into the backyards between the homes and the alley. Although dressed in a shirt and tie and in his late forties, Matt looks like an early millennium David Bowie, slender with straight, light-blond hair, highlighted from the California sun and parted perfectly down the middle.

I wave up at him, trying to get his attention. "I'm Nadina."

"Oh," he says, climbing down to hug me. "Am I late? I got distracted by these Mexican fan palms." He nods toward the yards. "Never seen them from this angle."

No problem at all, I tell him. The truth was that I loved meeting others at parks because if they ran into traffic, I'd be left with my book, free to appreciate the outdoors.

He types a note on his phone, then puts it away. When he smiles, his whole face lights up. "Are you ready for a long walk?" he asks, eyeing my sneakers. "There's something, in particular, I wanted to show you."

After four years as an urban forester, Matt has now held the position of public landscape manager for nearly five years, a platform from which he guides the future of the coastal city's development.

I had known about Santa Monica's tree inventories for almost as long as Matt had worked for the city. The system for taking stock of the cities' trees and replacing those that failed, which Matt and his predecessors had developed, was considered one of the best in the industry. Santa Monica's trees may not immediately come to mind when one thinks of the surfing town overtaken by Los Angeles. Still, as the tree inventories revealed, the city is home to a wide variety of species. The coastal environment, beloved for its scenic beaches and vibrant lifestyle, lends itself to trees that thrive in a warm, dry, and sunny Mediterranean-like climate. The tall and slender eucalyptus is a mainstay, swaying elegantly and adding vertical depth to the skyline. These trees, native to Australia, were brought to Southern California in the 1850s for their rapid growth and resistance to drought. Peruvian pepper trees, imported from South America's Andes Mountain range, dot the landscape, instantly recognizable by their delicate, scented leaves, droopy branches, and gnarled trunks. The iconic California live oak is also a common sight, with its twisted branches and lush green leaves that shelter all kinds of wildlife. The towering coral tree, with its striking red flowers, adds a touch of vibrancy year-round. And the city takes pride in the beloved jacaranda tree, which blooms in a breathtaking spectacle of violet in the spring. Along with the iconic palm trees that define the city in the popular imagination, these trees create a picturesque canopy that shades the streets and sidewalks below. Other West Coast cities, impressed by Santa Monica's success in increasing the city's street trees that continued growing in population, modeled their efforts after those Matt led.

But as Matt confessed to me, he wasn't as convinced that his celebrated tree inventory succeeded in protecting the city's forest. So one day he decided to set aside the database his employees painstakingly assembled and commission a study that would look at Santa Monica's canopy from an entirely different perspective—a bird's-eye view conducted via satellite imagery.

The results of the canopy assessment shocked many in the field. At the same time that many were cheering Santa Monica's urban forest program as a breakthrough model, the city was losing trees at a massive and unprecedented rate. As it turned out, the modest gains Matt and his colleagues secured on public lands were dwarfed by the massive destruction of trees on private tracts, where city workers couldn't readily observe them. The city's privately managed forest comprised nearly half of Santa Monica's trees, and it was shrinking year after year. From 2014 to 2020, the canopy on private land fell by almost thirty-three acres, resulting in a net loss of twenty-seven acres across the public and private forests.

Matt estimates the loss to be equivalent to 2,897 full-grown trees. "The problem," he says, "is the city only has enough space on public land to plant another 2,000 or so. We simply can't outplant private destruction." That reality had spurred him to take political action.

It came in the form of a city ordinance that he helped to champion in early 2020, which would allow for the regulation of trees on private property. Several city council members had expressed interest in it, preparing to bring it up for a vote. Then, the reform was put on hold as the COVID-19 pandemic descended and cities were overwhelmed with challenges to other services such as public health and schooling.

Matt's mission is to restore the momentum the proposed ordinance lost. "With the right data," he says, "we can prove that preserving private nature is essential."

The main cause of tree loss on private land in Santa Monica is the same one sparking urban deforestation across North America: development on residential single-family lots, typically to expand houses or add accessory dwelling units (ADUs). Tiny houses that offer affordable living space for extended families or additional rental income, ADUs have soared in popularity in recent years. And while the need for affordable housing is great, the increasingly bare satellite images Matt shows to city managers worldwide highlight better answers than this kind of low-density horizontal development.

Matt leads me down a street with impressive mansions set back from the road, each surrounded by tall wooden fences. I can see the floor-to-ceiling windows along the top floors of the homes, framed by a few crowns of trees but little else. I'm beginning to understand the challenge that Matt's surveyors faced before the switch to satellite.

He slows and catches my eyes as we approach an enormous flowering shrub. He brings a finger to his lips as if to hush the noise of the traffic passing us, then points through a wrought-iron gate that twists across a driveway. I can see a patch of ash trees staggered across a small hill through the gate. Below, three ground squirrels dart through the grass. One races up a trunk. A branch bends as a pigeon stops to rest.

"This is one of the city's largest private canopies," Matt whispers. A breeze shuffles the leaves on the ground. "I spotted it from space," he adds.

In recent months, Matt has begun working with a technology company called PlanIT Geo, which amongst other urban forestry services, compares satellite imagery every few weeks to pinpoint trees lost on private land. In Europe, where many cities prohibit private tree destruction without a permit, similar technologies to PlanIT Geo have quickly become a staple. In 2023, the government of the Flemish region of Belgium announced a new project in which every tree or hedge of at least three meters would be automatically registered using similar artificial intelligence technology, allowing officials to catch illegal fellers in the act.

Even as Matt uses the tool to collect the data necessary to make a case for a private tree ordinance in Santa Monica, he has no illusions about the legal challenges he would face when monitoring private compliance. The reason European cities have been able to pass private tree ordinances has to do with their legal structure. Many large cities in countries such as the U.K., the Netherlands, and Germany, have a ground-lease system. Essentially, the person buying a building or apartment owns the structure but not the ground on which it stands. Instead, they agree to lease the land for a certain period, paying their city a rental fee.

In the U.S., however, stronger private property rights are built into the Constitution. Even though courts have allowed certain kinds of government regulation of private land, like prohibiting contaminants that would poison aquifers, many other laws have been found unlawful, often for violating the Takings Clause of the Fifth Amendment.

In 2021, one Michigan town learned this lesson the hard way. Canton, a little-known township outside of Detroit, passed an ordinance requiring property owners to replace trees they removed from their property or pay into a tree fund. Canton was facing the same contagion of private tree destruction that Matt saw in Santa Monica, and a team of researchers had found that a tree fund could reverse the trend.

A local developer, F.P. Development, was one of the first entities fined under the law. Without applying for a permit, it had cleared fallen trees and debris from a county drainage ditch on its property, along with 14 landmark trees and 145 others. Canton ordered the company to replant 187 trees or deposit nearly $48,000 in the tree fund. Instead of paying, the developer sued.

It won in federal court, which found that the ordinance was an unconstitutional regulation of the landowner's property. But Canton, whose residents were alarmed by the city's disappearing trees, did not give up easily. The city appealed the ruling to the Sixth Circuit, which affirmed the lower court's decision. As each court saw it, the township could not provide evidence indicating that the company's tree removal had caused environmental degradation in the surrounding area.

The key to winning future legal challenges is obvious for Matt: cities must gather evidence linking tree removal to environmental damage and use it to design their laws.

Combining Dirk's LiDAR and PlantIT Geo's satellite analysis may offer the key to building that evidence. And fortunately, they are becoming more accessible. Apple's latest products now feature a built-in LiDAR scanner. It can create 3D representations of close-range objects up to five meters away. Similarly, Dirk's TreeTracker technology could be integrated into a mobile app where every homeowner, concerned neighbor, and watchful civil servant could easily—and accurately—measure a tree. That data could then be automatically input into i-Tree to determine the tree's "value" to the people around it. Savings in energy costs owing to the tree's shade, air-quality improvements, carbon sequestration, stormwater runoff absorption, and neighborhood beautification could all be accounted for. After all, if a tree doesn't retain stormwater, some form of gray infrastructure will have to do the job, and that has an identifiable price tag.

If it takes another fight in court to prove the constitutionality of private tree regulations paired with satellite-based monitoring systems, Matt isn't going to run from it. His British stiff upper lip and self-described "no-fucks-given" attitude make him an ideal candidate to lead this charge.

As we near the end of our walking tour of Santa Monica's arboreal wonders, I ask Matt how he keeps his spirits up. Now equipped with the tools to witness great losses, you'd think his work might tend to bring him down.

"A society grows great when old men plant trees in whose shade they shall never sit," he tells me, quoting a Greek proverb. "I'm not an old man just yet, but I will be when I retire from fighting for the trees we know future generations will depend on. If I need to be people's punching bag in the meantime, then so be it."

CHAPTER 3

Throwing Shade on Extreme Heat

When I feel the heat, I see the light.
—Everett Dirksen

I KNOCK URGENTLY AT the door. "Mrs. Temple?"

No answer.

I knock again.

"Mrs. Temple, are you in there?" I wipe the sweat from my brow and notice a small puddle of condensation beginning to take shape beside me. The box of Popsicles I brought to share with my ninety-seven-year-old neighbor is melting. It's July 29, 2019, and I haven't been able to sleep in days, kept up by a string of searing nights that we might have called "unseasonable" a few years ago. Lately, they've become all too common.

Today, it's over 101°F, and I'm worried about my vulnerable neighbor's well-being.

I knock again, and remembering Mrs. Temple's hearing problems, I call out, almost yelling: "Mrs. Temple?! Are you home?"

No answer.

I run down the stairwell of my Amsterdam townhouse to the door of the apartment two below mine, which I proceed to bang on so powerfully I fear I might blast it off its hinges. "It's Nadina."

Mrs. Temple's seventy-two-year-old niece, Maria, answers. "Yes?" She's panting as if she ran to get to the door.

"Mrs. Temple isn't answering her door. I'm really worried. Have you seen her?"

"She should be home . . ." Maria purses and then bites her lip. "She was this morning."

Maria's eyes fall on the waterlogged box of Popsicles in my hand. I offer her one. When I hand it to her, it feels like slush.

"Try it again in a bit, dear," she tells me. "She's probably just on the phone with a friend."

I walk up the stairs to my apartment and return the Popsicles to the freezer. Realizing I'm late for an appointment, I grab my keys, run down the stairs, and cross the street. I pass before two men spraying water onto one of Amsterdam's many bridges. It's a feeble attempt at restoring some kind of normal. By now, most of Amsterdam's shipping traffic has already stopped after strategies such as this couldn't keep the heat wave's temperatures from expanding the drawbridges' metal so much that they cannot be raised for passing ships. Canal traffic would be suspended for more than ten days.

The buckling asphalt of the street smells freshly laid, though in reality, the road hasn't been repaved in years. I glance at the tram station across the block and see that even that is closed due to the risk of derailment. The blistering sun and sweltering heat have no doubt started to warp and bend the tracks. It's a frightening collision of events, given public transportation's critical role in bringing vulnerable people like Mrs. Temple to cooling centers.

I reach for my iPhone to give Mrs. Temple another call. Then I remembered that I didn't have it—when I left it on the windowsill yesterday, it was scorched beyond repair.

Crossing Amsterdam's central park, Vondelpark, and making my way to the appointment, I spot Mrs. Temple. She reclines on a shady bench, reading the newspaper beneath a large oak tree.

"There you are!" I call out to her.

My adrenaline subsides. I'd heard on the news that morning that the Dutch Red Cross calls on family and acquaintances of vulnerable older adults to pay extra attention to them during the heat wave. Research by the organization shows that only one out of five Dutch people regularly ask older relatives if they need extra help because of the heat, even though every summer's grim statistics teach us that these relatives often do. Elderly women, in particular, showed worse health outcomes compared to men. In 2018, the European Union recorded 104,000 heat-related deaths among older people, over one-third of the global total. This year was looking far worse.

Mrs. Temple looks up. "Sweetie, hiiiii!" she says, moving her paper to the side and grasping my hand in both of hers like she always does. This time, her hands are clammy.

I sit beside her, forgetting for a moment that I'm late. "I hope you're being careful in this heat," I tell her. I give her my water bottle. "Take this," I say. "You need to drink more than you think in the heat."

She fixes her eyes on the small pond ahead of her and the patch of crisp, brown grass beyond it, where two dogs circle each other with uncharacteristic laziness. Nearby, two crows fight over a french fry, and a group of children dangles their toes in the water from atop a boulder.

"If I'm not home," she says, "you can find me here, cooling off."

Mrs. Temple and I are lucky, living across from Amsterdam's biggest—and in our opinion, prettiest—park. Most people suffering from heat waves have it even worse.

During July 2019, Amsterdam and the rest of Europe were experiencing the warmest summer on record. But recent trends suggest that we all need to get comfortable with heat waves that were unprecedented just a few years before. The record that the heat wave shattered had been set just a year earlier. In 2022, those heat records were again surpassed, with temperatures in the south of the Netherlands exceeding 104°F. It was the third time reaching 100°F in recorded history and the third time in the past three years.

And the Netherlands is far from the only country that has found itself the victim of an increasingly deadly summer climate. In July 2019, France measured its highest temperature ever—112°F—a record again shattered later that day when the southern municipality of Gallargues-le-Montueux rose to nearly 115°F. The latter broke Europe's previous high, set during the 2003 European heat wave that had taken an estimated 70,000 lives. At the end of 2018, each of Europe's five hottest summers in the past five hundred years had occurred in the prior fifteen. Then, 2019 proved to be the continent's hottest yet, then 2022. That is, until 2023.

The summer of 2023 unleashed a global heat wave of unprecedented intensity. It began in the southern U.S., where June set a new record as Earth's hottest month since recordkeeping began in 1850, according to the National Oceanic and Atmospheric Administration (NOAA). Phoenix, Arizona, for example, endured over four weeks of temperatures above 110 degrees Fahrenheit, claiming twelve lives in Maricopa County alone. The intense heat put enormous stress on both human health and infrastructure, with more than 100 million people in the U.S. under excessive heat warnings and advisories.

As we moved into the heart of summer, the U.K. faced an unusual marine heat wave, Siberia saw temperatures surpassing 100°F, California's Death Valley, reached a scorching 128°F (a rarity on Earth), and China registered record-breaking hot days. By mid-July, southern and eastern Europe faced extreme temperatures reaching 118°F in Milan, Italy, where a road sign worker succumbed to the searing heat. In Athens, temperatures soared over 100°F, leading to the closure of the iconic Acropolis to ensure visitor safety. The heat also sparked wildfires, causing numerous people to evacuate resort towns southeast of Athens.

This wave of extreme weather is largely attributed to the Charon anticyclone, a sweltering "heat storm" originating from North Africa. This follows hot on the heels of the Cerberus weather system that pummeled the region just a week prior, heralding a summer that continues to test Europe's resilience to extreme weather.

Heat, especially for vulnerable populations, is a grave issue. According to a report by the *Lancet*, one of the world's top medical journals, Europe's aging and city-dominated populations were to blame for Europe's increasingly fatal summers. Commenting on the article, the *Lancet*'s editor-in-chief wrote, "The time [has come] for us all to take the environmental determinants of health more seriously—we must address the climate emergency, protect biodiversity, and strengthen the natural systems on which our civilization depends." In tallying the body count, the medical community could no longer ignore a simple truth: that urban green space is a matter of life and death.

Most northern and western European cities aren't built for the stifling periods that have become a summer tradition. Historically, with average July temperatures of less than 72°F, there was little reason to design for them. Take my apartment in the center of Amsterdam. Built in 1911, it betrayed the priorities of home builders for much of the past two centuries when they sought to fit as many exterior windows as possible. The strategy aimed to bring in light, relieving what we have since come to call seasonal affective disorder, which affects up to 8 percent of all Europeans annually. My apartment, which I now begrudgingly call "the greenhouse" in the summer months, excels at what it was designed to do: pull heat in—not let it out.

France's 2022 record heat wave brought the nation into a once-exclusive club of southern European countries where the mercury has broken 113°F, including Bulgaria, Portugal, Italy, Spain, Greece, and North Macedonia. Once known as the only country in Europe that is both a northern and southern power, France has paved the way for other northern European countries to join the heat club.

As Greece discovered in 2018, extreme heat can trigger other cascading natural disasters. The prolonged periods of excessive heat burns humidity out of the air before it becomes rainfall, exacerbating drought. Through this process, they create the ideal conditions for wildfires. Such was the case when a months-long drought fueled by record heat receded into a windy autumn in which hundreds of wildfires scorched Greece. The devastation came only three years after fires in coastal Attica, the region that includes Athens, became the world's second-deadliest fire event of the century, killing more than a hundred people.

Newer members of the heat club, including the U.K., are facing unprecedented "grass fires," where scorching and dry weather causes prairies and parks to burn quickly when exposed to even the smallest sparks. In June 2022, London firefighters tackled more than eight hundred grass and open-land blazes. When reflecting on the fire service, London's mayor told the BBC it was the brigade's busiest day since World War II.

Elsewhere around the world, more cities are punching their membership cards—and the threat we've imposed on nature has begun to threaten us. Research by the U.S. Forest Service shows the country is losing 36 million trees annually in urban communities. Nearly half of the areas where we've chosen to build new houses, schools, and institutions used to have trees. Even if we wanted to replace each of the 360 million trees that the U.S. lost in the last decade, we're rapidly running out of space to grow them. In all these areas where nature has disappeared, people feel it daily. Fewer trees filter the air. Fewer wetlands and marshes clean the water and protect communities from floods. Fewer parks offer children a place to play and adults to unwind. And fewer public spaces invite people to forge community and build solidarity. Many feel the loss of greenery most sharply during the increasingly severe summer heat. Since trees can lower summer daytime temperatures by as much as 10°F, recent urban deforestation already leads to an additional 1,200 heat-related deaths annually and more than 100,000 extra doctor visits for heat-related symptoms, according to a study by the Nature Conservancy.

In June 2021, the Pacific Northwest of the U.S. and southwestern

Canada—a region that supports some 13 million people and is known for rainy, mild weather—experienced its worst heat wave on record, right on the heels of record-breaking temperatures across the American southwest. That "heat dome" lasted five long days, leaving over one billion sea creatures on the coasts cooked to death. Dead mollusks and clams, parched rockweed, dehydrated starfish, and baked barnacles were among the creatures that littered beaches along the Pacific coast.

For these creatures, already immersed in water—nature's best cooling mechanism—there was nothing to save them. A starfish's family, friends, and neighbors were equally at risk of succumbing to the heat. With humans, however, that isn't the case. Heat waves disproportionately affect those in cities that lack access to air-conditioning and cooling green areas. For people like Mrs. Temple and me, access to green space made all the difference.

The instinct to seek cool, natural environments was not some quirk of Mrs. Temple's. It's how humans stayed cool throughout our evolution and the solution to our current battles with extreme heat. The question facing cities, confronted with rising temperatures, is how they can feasibly redesign themselves to allow inhabitants to stay cool.

The answer must address the immense challenges of traditional urban design, which serve to amplify the effects of extreme heat. The concentration of buildings and roads in compact urban areas creates what scientists call an urban heat island effect, in which cities experience higher temperatures than their surrounding rural areas. Common materials used in urban infrastructure, such as concrete and asphalt, absorb sunlight and reflect heat. Poorly designed buildings and streetscapes with abundant motor vehicles can also impede natural airflow, trapping heat and pollutants in urban areas. These factors contribute to the scorching summer in cities that are only worsening due to climate change.

Reducing the impact of extreme heat in cities requires a thoughtful design that incorporates green spaces and vegetation, which play a crucial role in mitigating urban heat islands through two primary mechanisms: shade and evapotranspiration. Trees and plants provide much-needed shade, lowering the surrounding temperature by absorbing incoming solar radiation and effectively reducing reflective heat. In addition, the process of evapotranspiration, whereby plants absorb water from the soil and release moisture into the air, creates a cooling effect. Research shows that cities that invest in green spaces and vegetation significantly reduce the urban heat island effect, enhancing the well-being of their citizens and the natural environment.

As climate change heats up the world's summers, cities' efforts to counteract the urban heat island have taken on a new, existential importance. If they cannot provide a comfortable habitat during the warmest times of the year, residents may be forced to flee to greener—and cooler—pastures. The severity of this crisis cannot be overlooked because it can potentially leave our homes—and the urban engines of the modern world—desolate and uninhabitable.

Yet despite several degrees of rising global temperatures already inevitable, there is still hope to save our cities. As several groundbreaking researchers have recently discovered, we can live cooler by bringing the natural environment closer to our buildings, all while saving on energy costs.

Dr. Vivek Shandas, professor of climate adaptation at Portland State University (PSU) and founder of Climate Adaptation Planning + Strategies (CAPA), is one the leaders of this group. He studies the disturbing cost of our unequal access to natural cooling resources.

To better understand how those impacts played out in the 2021 "heat dome," Vivek agreed to take me to ground zero.

When he finds me outside his office, he gives me a bear hug. Though we don't know each other well, we have bonded at nature conferences over our love for the urban forest, and I always get the sense that he is the kind of person who makes everyone feel comfortable. In many ways, Vivek embodies Portland: a politically engaged hobbyist hiker with a laid-back smile and a deep commitment to nature conservation. Today, he's wearing a green-gray checkered jacket with slim jeans and worn sneakers caked with mud, his face tan from this summer's measurement campaigns. If it weren't for the faint salt and pepper on the sides of his faded undercut, you'd never know he'd just turned fifty.

We're in Portland's Central Eastside Industrial District, where not a tree can be seen. The industry in the neighborhood's name left decades ago, making way for start-ups, stores, and lofts that have filled the former warehouses.

"How is Amsterdam faring," he asks, knowing it has already been a tough summer.

"We're managing," I say out of pride, realizing it's a lie. "You?"

"Us, too," he says. "I guess." I can see the pain behind his eyes. "It's been a lot."

Being outside, no matter how bad the weather, is how Vivek deals with adversity. As his Portland neighbors retreated to air-conditioning to wait out the heat dome—at least those lucky enough to have AC—he hit the streets, taking the temperature of dozens of blocks. Nature has been his refuge for as long as he can remember. Just ten years old when his family moved from Bengaluru, India, to Santa Rosa, a small city in Northern California, he found that he was one of the only Indian kids at his elementary school. He felt isolated, thousands of miles from the boys he grew up playing cards with. But climbing up muddy trails at the foot of the Sonoma Mountains, Vivek discovered that in nature he felt at home.

Unsurprisingly, despite coming from a family of engineers, Vivek decided to pursue environmental sciences. Fresh out of school, he worked on the Landels-Hill Big Creek Reserve on California's scenic Big Sur Coast, building out the university's herbarium. His $100 monthly stipend covered rice, beans, and all the other essentials he needed while living out of his tent. It was one of the happiest times of his life.

Vivek's experience at the reserve inspired him to share the power of nature with others. Soon, he found himself teaching "outdoors school" in Oregon. The state mandates that middle schoolers spend a week camping and learning in the world's greatest classroom: nature. Vivek understood why experiencing what he had would be valuable for kids. But he wasn't prepared for how the city kids he taught would change his perspective.

For several years he tried to bring the immersive education of nature to children in the city. He developed, among many other things, a curriculum for teaching how to test local water quality, which the San José Children's Discovery Museum translated into Spanish and still uses today. Later, Vivek became a professor at PSU and shifted his focus to studying heat across urban neighborhoods and the people it harms the most.

As we drive down one of Portland's wide streets, Vivek lowers the windows to welcome in the breeze of a green neighborhood we entered a few minutes ago. "We're about to be in East Portland," he says, reminding me to pay attention to the streetscape. It rapidly changes from lush, tree-lined streets to what Vivek calls "anywhere, America," a mosaic of strip malls, parking lots, and asphalt roads stretching for miles. We pass a cemetery on my right-hand side, strangely a welcome patch of life in a pallet of gray.

"What's weird," Vivek begins, "is that we used to understand urban greenery as a gradual buffer to heat." But with more extreme heat waves, he tells me that's changed. He gestures toward the cemetery. "On an eighty-five-degree day, you'll feel the cooling incrementally as you walk closer and closer to that grass. But on a hundred-degree day, there's no buffer—you'll immediately feel the heat stepping off the green lot." That may not seem particularly significant, but Vivek claims that the hotter the day, the harder nature has to work to combat the temperatures. The principle that this demonstrates will have unprecedented impacts on the neighborhoods we build. Without a green intervention, they'll continue to react to temperatures we've never seen in a vicious cycle. To properly mitigate against temperatures as high as they were in Portland, we need more green everywhere.

As Vivek and I drive deeper into Southeast Portland, a historically disinvested area, the strip malls increase. We stop at a traffic light where Vivek points out a "massive missed opportunity," a newly constructed apartment building. The top-story windows don't open, and the bottom ones only crack open a few centimeters. The units don't have AC, and there's no greenery in sight. "It's a greenhouse in there. That building will kill people, and they don't even know it."

The observation hangs heavy in the air.

A few blocks later, I break the silence. "That building—what's the best way to—"

"Cross ventilation instead of boxed-in walls, green walls, green roof," Vivek rattles off. "There needs to be a median strip to allow canopy growth."

The problems with that building are the same ones he sees everywhere: we're still designing for yesterday's conditions, putting up structures that we expect to house people in fifty, seventy-five, a hundred years—except by then, they'll be completely uninhabitable.

We drive five more minutes past Arby's, Staples, Lowes, and more edge-to-edge housing developments. Then we roll to a stop.

"Is this it?" I ask.

Vivek nods and reaches for the car door. "Heat dome ground zero."

Last summer, this zip code experienced the highest death toll. Vivek notices me looking across the eight lanes of the Interstate 205 overpass some 100 feet ahead of us, where there's a small tent city even on this unseasonably hot October afternoon. The overpass is the only infrastructure providing any shade in the vicinity. "You see those boulders?" he asks.

I notice them spaced several feet apart, lining both sides of the underpass. He explained they were the city's response to the local homeless population, deterring people from camping. That move turned out to be deadly. "Some people took their tents out from under the highway," he said, "on a day I measured the temperature here at 167." He takes a few steps toward the underpass, then stops, resting a hand on the concrete wall that lines the sidewalk, like he is catching his breath.

"Right here," he says. "This is where one of the people forced from the underpass died."

Despite the seeming inhumanity of it all, Vivek doesn't blame the city. "The city wasn't thinking about the heat when they put those boulders down." He sighs. "But that's exactly the problem—they're never thinking about heat."

Vivek's personal and professional lives collided when the heat dome descended on his city in 2021 during what many Americans across the West called an "apocalyptic" heat wave. Ten days before the heat dome, he had seen its first signs, as his weather model projected it to climb to 114°F. Perhaps out of wishful thinking, he convinced himself it couldn't be. The daily highs exceeded even July averages for some of the hottest cities in America, like Phoenix, Arizona, and Las Vegas, Nevada. Even the evangelist of remedying the city's massive heat problem thought it impossible, chalking it up to a modeling error. So he checked the National Weather Service's forecast. Then he looked at his own the next day, the next day, and the day after. Five days out from the predicted heat wave, when the accuracy of Vivek's model began to improve exponentially, the temperatures had only increased.

Vivek brings me to a bench, telling me what he did that week—from the first blistering hour to the thick clouds that finally swept the heat away. "My first reaction was—" He gulps. "That people are going to die. Second was that we only had five days to get help."

Decades of heat research flashed before his eyes as he considered what could be done on such short notice. He had interviewed people living in tents, trailer-park homes, and cramped apartments with multigenerational families. "It felt like everything I had written about, everything I had predicted, the social vulnerabilities, the historical land use, the blasé attitude toward heat in the Pacific Northwest. It was all swarming before me, and I was completely helpless."

Ultimately, the help that Vivek hoped for never came. Only a few experts seemed to express alarm three days before the heat dome. There had been no warning from the National Weather Service and no texts or

phone calls from the city. He started asking those he knew in the local
government about their plans.

"Finally," Vivek says, "an advisory went out the day before it started."
As the sun rose the next day, the National Weather Service issued another, telling residents to try and stay cool. On the news, they said it
would be 112°F, but Vivek knew better than that: his research had shown
him that heat is unequally distributed across the city. He couldn't answer the question of how variable it would be at temperatures the region
had never seen. Armed with a thermometer and an infrared camera attached to his smartphone, Vivek and his then eleven-year-old son, Suhail, were determined to find out.

It was important for Vivek to take his son out with him, not only
because Saturday was the day of the week that Vivek always set aside to
spend time with him, but because this was an opportunity to share an
important life lesson. "The local meteorologists were calling this a one-
hundred-year event," Vivek says, "and I told Suhail not to think about
this day as the hottest it's been in the last hundred years but as one of the
cooler July days that he and his children would experience in the *next*
hundred years." He smiled, remembering how Suhail's eyes had glazed
over. "Suuure, Dad," Suhail had said.

Whether or not the message had hit home, the newly minted family
team started their measurements on the corner where Vivek and I now
stood. It was a typical North American development, where building
walls met sidewalk and sidewalk met asphalt road, a collision of concrete reflected down the four streets that sprawled in each direction. On
this warm afternoon, the sheer face of the building wall offered little
shade. There wasn't a tree, patch of grass, or vine of ivy in sight. I had
a hard time not picturing the man who'd died on this very sidewalk,
slumped against the radiating wall, seeking refuge where none existed.

"It reached 124 degrees," Vivek says, breaking my lurid daydream.
"And it didn't drop below that." Vivek told me how he'd coaxed Suhail
to walk another couple of blocks—quick to confirm Suhail was well hy-
drated and "lunched up"—to see if the temperature would fall. It was

three P.M., and the sun was starting its slow descent. But on that day, it only grew hotter.

After taking measurements for another thirty minutes, the father-son duo had proceeded back home. Vivek and I got back in the car and set off in the same direction he'd traveled with his son, keeping our windows down. A few minutes later, I noticed a change in our surroundings. "Where are we?" I asked.

It was Northwest Portland, Vivek told me, a neighborhood that looks—and feels—completely different. Lush pin oaks line the narrow streets; the homes set back several feet from the sidewalk. In each traffic median, a flourishing ecosystem of shrubbery blossoms. Vivek pulls his Prius to the side of the road, across from a man raking leaves on one of the medians.

On this block, Suhail had measured the temperature at 99°F, 25 degrees lower than the reading he'd taken just a few miles within the same hour. Astonished by the difference, he and Vivek canvassed the neighborhood to confirm the measurement wasn't anomalous. "Not once did the thermometer break one hundred," Vivek tells me.

Always the scientist, he's quick to point out slight topography differences between this hilly neighborhood and the urban desert we'd just left. But he insists they are insufficient to account for the temperature difference.

So, what is it?

"Obviously, there's the shading aspect," he begins, "but the real magic is something else." When leaves are exposed to the sun's rays, he says, water evaporates off them, cooling the surrounding air. And it doesn't have to be trees. Smaller plants also evapotranspire, releasing water vapor into the atmosphere, which carries away heat and reduces the overall temperature. While trees are often considered the primary solution for cooling urban areas, scientists find that smaller plants like shrubs and grasses can also contribute, working together to cool our cities.

This effect explains how on the second day of the heat dome, Vivek recorded the surface temperature of an ivy-covered green wall beside his industrial office building at 119°F, while right next to it, a wall that lacked ivy registered at 157°F, an astonishing 40-degree difference. Both walls were directly in the sun.

Inside homes, the contrast is even more stunning. A study from the University of Reading in the U.K. found ivy to be the most effective plant cover for cooling buildings and reducing humidity. Not only does it provide the best summer cooling for buildings, but it also keeps them warmer in the winter. Another study found that houses insulated with a vertical wall with plants had 30 percent less heat loss when compared to a room without such a wall, slashing a home's energy bills.

When Vivek returned home that evening, drenched with the sweat of the day's measurements, he input all the data into a spreadsheet. Seeing it side by side blew him away. Neighborhoods lacking plant coverings—typically the poorer ones—reached temperatures as high as 124°F, hotter than the highest recorded temperatures in Las Vegas, a thousand miles to the south. In stark contrast, greener and wealthier Portland neighborhoods averaged about 98°F.

The extreme contrast in the temperature data that he proceeded to publish, coupled with the high mortality rate the city announced, finally had people paying attention. Within a month, the *New York Times*, *Washington Post*, and *The New Yorker* lit up his phone. It was bittersweet for Vivek, who'd been laughed out of the room for suggesting Portland could get so hot just seven years prior.

"I was in front of all the right people then," Vivek recalls, "Oregon Health Authority, the city of Portland, Metro Regional Government, local health agencies . . ." he trails off. I swear I see his face wince in pain.

"We could've saved many lives," he begins in a higher pitch, "but it wasn't a priority for them."

A few weeks before the heat came, Vivek had been presenting the results of a National Science Foundation–backed heat study, the first to project that these kinds of scorchers would come within the decade. The study laid bare the vulnerabilities of Portland and other neighboring cities, whose built environment and lack of air-conditioning put residents at grave risk. Having grown up in the searing climate of Bengaluru, he understood the impact heat waves had on the most vulnerable. But as he waited at the end of his presentation for questions that never came, he wondered if others could imagine the destruction he feared. "There was a belief," he tells me, "one that was pervasive and deeply held—especially for those who'd lived here for decades—that, no matter what any model said, Portland just couldn't get *that* hot."

The humiliation of so many similarly tepid receptions to his work might have been enough to send most scientists into hiding. Vivek, on the other hand, simply wanted to learn how to move forward, blaming himself for failing to inspire the fear that would be rational. He concluded he wasn't doing a good enough job of communicating the looming, sinister threat. If people were going to take this seriously, he decided, they needed to be able to see on a hyper-local level what heat could do to people. At the time, heat maps were primarily based on land-surface temperature measured by satellites. Like thermometers in the sky, satellites measure part of the atmosphere up to 10 km above our heads, known as the troposphere, inferring ground temperatures from that data. While the method has important benefits—namely, the fact that it's replicable and scalable worldwide—its measurements are coarse. They perceive almost nothing about conditions varying neighborhood by neighborhood, block by block, home by home, the measurements that Vivek was convinced were necessary to save lives.

As the heat in the Pacific Northwest receded, its grim fatality was undeniable. Across major cities such as Portland, Seattle, and Vancouver, hundreds died, and thousands more were hospitalized with heat stroke, heat exhaustion, and dehydration. By the end of summer 2021, Americans had confronted a reality that many Europeans were still struggling to grasp: during a heat wave, every degree increase in Fahrenheit temperature can raise the risk of dying by 2.5 percent.

Research has shown that if the urban forests in the U.S. disappeared tomorrow, there would be twice as many deaths due to heat. And by contrast, merely covering 30 percent of cities with trees would cut heat-related deaths by a third. Yet despite the increasing human toll of our bare surroundings, most municipalities have invested far too little to understand how streets can offer enough natural shade to combat heat deaths and which improvements urban planners must prioritize.

In my early research, I found the lack of investment in heat-fighting green infrastructure hard to make sense of. Compared to large-scale infrastructure projects, it was obvious that urban greening was readily achievable, and the economics, medical literature, and climate research all pointed to its life-saving potential. But when I launched my first discussions with my local policymakers in Amsterdam in 2015, I had my first taste of the challenge Vivek encountered daily. The city planners didn't know where to start.

"We don't have the resources to send people walking around our streets, recording the sun" was a sentiment I heard more than once.

But what if we did? Imagine a city that could access a true map of its neighborhoods—not just its streets but the buildings, structures, and greenery surrounding them. What if city officials could identify which roads provided adequate shade and which needed more plants to cool them? The result would be a cheap, verdant, and highly utilitarian streetscape.

More than a year later, I met someone who showed me this was more than a dream.

I'd just moved to Cambridge, Massachusetts, to join MIT's Senseable City Lab, a research group that studies the interface between cities, people, and technologies. There, I was to embark on six months of Ph.D. fieldwork, sampling and analyzing microbial activity in the soil of the city's red maple trees. On my first day, I met Dr. Fábio Duarte, who would become one of my supervisors. Though I didn't realize it, I'd followed him to MIT, having just moved from Amsterdam, where he'd spent the summer doing fieldwork. We bonded over our shared appreciation for that city's greenery, which had kept us cool through yet another sweltering summer.

Having grown up in Brazil, witnessing São Paulo's slums, called favelas, Fábio understood the dangerous health impacts of extreme heat all too well. The new climate he was enduring in Massachusetts and the Netherlands felt like the "Indian summers" he experienced as a kid—only more frequent, intense, and longer lasting. Now a senior researcher at the Lab, he was working to help his onetime neighbors.

Since all kinds of greenery impact a neighborhood's ability to avoid dangerous heat levels, Fábio focused his first project on measuring plant growth.

On my first day in the office, he set aside the morning to show me the ropes. Before I could even finish the university's mandatory IT training, he hurriedly called me over to his computer monitor to show me how he and his team quantified greenery at the street level. Fábio could easily be mistaken for an enigmatic artist: he's slender, wears thick-framed specs, and his signature attire is black turtlenecks. He said that I, the Lab's first "tree person" in its nearly twenty-year history, ought to appreciate it.

The project he began to explain had started small, with ten cities, including Paris, Tel Aviv, and Boston, applying a technique known as computer vision. It enables computers to see, identify, and process images, like the human brain but on a huge scale. Fábio has another name for it: Treepedia.

"The technique is not new," he told me as if he were offended that others failed to use it while tens of thousands die yearly of heat stroke and related illnesses. "People had already trained computer algorithms to use it for much more trivial purposes—to identify cars, store facades, utility lines, and other things from Google Street View imagery. For some reason, nobody had done it for greenery."

That oversight contained an important lesson, Fábio thought: Sometimes we do not solve problems because of a lack of data or technology.

Sometimes we overlook something simply because we fail to pay more attention.

While Treepedia's website, which has now quantified over thirty cities, invites users to contrast street vegetation levels between cities, Fábio warns against participating in what he calls the "green Olympics." Every city exists in its own unique climate, so direct comparisons are useless. But comparisons *within* a municipality—neighborhood by neighborhood, block by block, plot by plot—hold the secrets for maximizing cooling benefits.

In 2017, Fábio persuaded São Paolo to try out his technology. The results showed huge disparities in greenery, with some affluent streets containing more than four times the city's average. Fábio understood that his technology could ensure that the city's future leaders pay attention to what their predecessors ignored. Yet despite the curiosity of São Paolo's leaders about their home's greenery, they didn't feel they could justify investing millions of dollars in plants when so many in the city suffered from hunger, a lack of running water, and other acute ills of poverty. The roadblock, Fábio realized, was that Treepedia couldn't demonstrate the life-saving impact of greenery through its cooling effects. That would require technology that showed indisputable proof of how barren streets become unbearable. In other words, Treepedia needed Vivek.

On that suffocating, fateful first evening of the heat dome in Portland, with Suhail tucked into bed, Vivek went back into the sweltering darkness, spending all night meticulously taking measurements. He finished in outer East Portland, one of the city's poorest neighborhoods. His data showed that the asphalt that characterized that area was an incredibly proficient insulator, keeping the area warm throughout the night. While the daytime deaths were alarming, the hidden deaths at night worried Vivek the most. Without relief from the heat at night, heat stress builds, increasing the risk of heat illnesses.

He worked through the dawn, intent on getting the granular data he needed. When Vivek got home, he made breakfast for his just-awoken tween. He promised Suhail he would find a better way forward before the next heat wave.

Two years prior, Vivek had founded CAPA and launched its first initiative: a national community-based heat mapping campaign called Heat Watch that trains volunteers to collect heat data. While Vivek loved conducting his field research, he knew it moved too slowly. People across the globe needed the temperature-monitoring equipment he'd spent six years perfecting and the data it unlocked.

Vivek thought he would sleep better with the technology now out of his hands, and for a while, he was right. But after the heat dome, that was no longer the case. He'd find himself lying in bed for hours, ridden with guilt that, as the resident "heat expert," he'd failed his home city in its crisis.

I wonder if the pain he experienced also related to another feeling, something more universal. In 2022, a landmark study found that the heat dome that also scorched British Columbia across the Canadian border led to a 13 percent rise in anxiety over the effects of climate change. It offered the first hints of the psychological impacts that may become ubiquitous as heat waves spread across the globe.

But whatever the source of his malaise, Vivek knew how he could recover: he took a sabbatical from PSU to focus on CAPA full-time, doubling down on his efforts to cool the places where we live.

These days, Heat Watch mounts the sensors that Vivek spent years perfecting onto volunteers' cars and bikes, asking them to follow a pre-determined route on a predetermined day. The trip collects data, which is later transformed into heat maps that inform urban policy. Since 2019, Vivek and the Heat Watch team have mapped over seventy-five cities. The method is so scientifically robust that in 2021, on the tail end of the heat dome, the NOAA funded Heat Watch to map over forty communities each year.

Neighboring North Carolina cities, Durham and Raleigh, were two grant-funded regions where volunteers drove, biked, and walked with sensors, collecting nearly 100,000 data points. The resulting heat maps revealed startling facts about the distribution of urban heat. Before Heat Watch, Durham and Raleigh's officials believed the hottest and coolest parts of the cities differed by 6°F. But Heat Watch exposed a much steeper margin: 9.6°F degrees in Raleigh and 10.4°F in Durham. The high-resolution maps showed that extreme heat is exacerbated by the presence of concrete and the absence of trees, leading to dramatic temperature differences in buildings across the street from one another.

Building from Heat Watch's data, Durham and Raleigh released new mitigation plans incorporating green roofs, porous pavements, and vegetation.

Such breakthroughs at the municipal government level excite Vivek, but not as much as when he sees children make sense of Heat Watch's insights for the first time. During a project visit in Richmond, Virginia, he met a high school student fascinated by the maps. "Wait . . ." the boy said, studying the map of his neighborhood, "this means that my friend's house is twenty degrees cooler than mine?" He pointed to the street separating them, half rendered in green, the other half orange.

Vivek later found out that the teenager had planned to drop out of school to make time for a retail job. But according to the supervisor managing Richmond's Heat Watch program, the map had ignited in him a passion for historical land use, development regulations, and green space policy. Not only did he finish high school, he pursued environmental sciences at the University of Richmond. "He's almost done," Vivek says, turning from me to wipe a tear away with his checkered jacket sleeve. "I can't wait to hire him," he says.

Although these kinds of stories cannot be quantified into the impact that Vivek can identify in the grant application, they have helped him sleep better.

Still, he strives to do more. For his final act in the field, Vivek confessed that he wants to prove the science behind the correlations between heat and street greenery that he's spent decades observing. Satellite imagery is a natural choice for such an analysis. Yet perhaps more than anyone else, he's attuned to its limitations—namely, its inability to capture greenery beneath the canopy and its effect on local, ground-level temperatures. Confronted with those shortcomings, Vivek is exploring how Treepedia could plug the holes in the data. Combining his heat sensors with Google Street View imagery, he could create fortified maps that identify where to plant, grow, and nurture greenery to build the livable communities of the future.

Having those tools at their fingertips may seem far off for many city planners. But one city is proving it may be feasible in the next few years. Unsurprisingly, it's in the heart of one of the countries hardest hit by the heat waves.

Paris is a tightly packed mass of metal, concrete, and asphalt that work together to exacerbate the urban heat island effect. Paris scored the lowest of the thirty-plus cities analyzed on Treepedia, with a staggering 8.8 percent of street tree canopy coverage. To make matters worse, Paris is one of the densest cities on Earth, with more than 20,000 inhabitants per square kilometer. The analysis hit a nerve for the city's deputy mayor, who felt Treepedia unfairly excluded parks that are invisible to Google Street View. But satellite imagery confirms Paris lacks greenery overall, with only 5.8 square meters of green space per inhabitant, well below London's 45 square meters per inhabitant and staggeringly distant from Rome's 321.

Despite her deputy mayor's embarrassment, Paris mayor Anne Hidalgo swallowed her pride to roll up her sleeves and combat extreme heat. In 2020, coming off the heels of yet another disastrous heat wave, Mayor Hidalgo proposed an ambitious plan to transform 50 percent of the city's surfaces from paved to vegetated and grow five "micro-forests" next to landmark locations. Springing in front of Paris's city hall, the venerable Palais Garnier opera house, the Gare de Lyon, and along the Seine quayside, these forests will forever change the city's architectural face. In addition, the mayor has pledged to plant 170,000 trees. And where there is no room for them, the city will create a hundred "oasis streets," alive with plant-covered walls, to cool streets down during the hottest summer days. These oases will be located near schools and in areas Treepedia identified as lacking green spaces.

If Paris, a city inextricably tied to its history, can evolve to combat the current heat crisis, then there isn't a place on Earth where natural shade should be out of reach.

However, before achieving a cooler future, we must overcome a stumbling block that has derailed years of attempts to grow urban vegetation: the wildfire risks posed by the wrong plants in the wrong places. After all, greenery, what we need to beat the heat, can become tinder for wildfires. Thus, to survive rising temperatures and unleash nature's astonishing benefits, we must unravel old assumptions about how we grow the wild.

CHAPTER 4

Spreading Like Wildfire

Two hands cut down a few trees,
but one match clears a whole forest.
—Matshona Dhliwayo

THE SIGN WELCOMES ME as I drive into Paradise, California: "Rebuilding the Ridge." Held up by two freshly cut two-by-fours, still golden from the lack of rain, the sign seemed to embody its message. Paradise was a community of survivors determined to rebuild.

The damage from the 2018 wildfire is everywhere, even four years later. The Camp Fire, named after its place of origin twenty-some miles away on Camp Creek Road, instantly became known as California's deadliest and most destructive wildfire. This state was already known for its devastating blazes. Later that year, when insured losses were finalized, it was deemed the most expensive natural disaster in the world.

The firestorm flattened 95 percent of Paradise and most of its surrounding communities, obliterating 18,800 structures, most of them homes. Eighty-five lives were cut short, twelve civilians and five firefighters were injured, and countless Californians became homeless overnight. What used to be a city of 26,800 lost more than 90 percent of its residents, forcing California's governor to reclassify it as a rural area.

Even those lucky enough to survive and have the money to re-build often didn't want to. The memories that the charred landscape evoked—of family businesses vaporized, of friends and family who were no more—were too traumatic. That was the case for Zeke Lunder, a wildfire analyst and mapping expert from a couple of towns over who had shown me the path of the fire across Northern California's Butte County. "You'll need to go to Paradise alone," he said. "I can't bear to see it anymore."

It is hard to imagine the vibrant community Paradise once was. But Magalia, its smaller neighbor five miles due north, offers a rough approx-imation. A few streets untouched by the fire remain: tree-lined avenues shaded by thick canopies and buttressed with newly constructed family homes. A couple walking their dog waves at me as I drive by slowly. Back in Paradise, driving through its streets by myself, the scene is so disturb-ing that I don't want to photograph it. I feel like a voyeur. As I drive off, I remember what Zeke told me and reach for my camera: "People need to know what's at stake."

The first thing you notice driving into Paradise is the scores of road and utility workers. No aboveground power lines remain on the Sky-way, the main commercial strip running through the heart of Paradise and Magalia. A dark line of fresh soil betrays that they're now buried in trenches that line the road.

The fire started in early November 2018, though the conditions that facilitated it had been building for months. Paradise, which typically re-ceives more than five inches of rain in the first eight months of the year, had only gotten one thirty-fifth of that. As autumn crept forward, the humidity dropped, and the hot Diablo winds howled through the dusty valley. Around 6:30 A.M. on November 8, one spark was all it took for the gusts to ignite a sea of fuel, launching a fast-moving firestorm. A six-month investigation led by the California Department of Forestry and Fire Protection (Cal Fire) concluded the fire was caused by a single failed component on a Pacific Gas and Electric Company (PG&E) elec-trical transmission tower, causing a short circuit and throwing sparks into the dry brush below—with a predictable outcome. Less than six hours after it started, the fire had ravaged Butte County's ridge-top towns, leaving charred ruins in its wake. It was the disaster of a lifetime, whose acceleration blindsided even those who had warned about such a storm for decades.

As a result of that speed, many residents perished in their homes, others attempted to escape in their cars only to find themselves trapped by flames. Some abandoned their cars, dodging the flaming embers on foot but suffocating as the smoke overwhelmed them.

Today, the wounds of that day still gape—in the form of hillsides, barren except for the pioneer species climbing up from the valleys, whose spicy scent has replaced the lingering stench of smoke in the trauma of survivors and first responders that seems to ooze from the rubble left behind.

It's 8:30 A.M. on a Tuesday, but no children are walking to school and there is no commuter traffic. The bakery down the street has boarded-up windows, and a "Closed" sign hangs crookedly across the blackened door. Turning onto a narrow residential street, I pass a gas station that looks like the siding has melted off, a burned-out car sitting in the corner. The dusty road gives way to rows upon rows of lots where houses used to sit and children used to play, aluminum fences marking the plots where families used to garden. The few with insurance covering fire damage lived out of trailers where their houses used to be. Most have put up a "For Sale" sign, surrounded by scattered stumps and blackened tree trunks.

The Camp Fire may have been the deadliest and most destructive in California's history, but it was far from the only one. And trends suggest it will not be the worst. Of California's 32.1 million acres of forest land, 6.6 percent burned from 2002 to 2011. In the following decade, that figure quadrupled to 24.7 percent.

Before it passed to the Camp Fire, the title of the most destructive wildfire in California history was the 2017 Tubbs Fire, which destroyed at least 5,600 homes and structures, and took twenty-two lives. Despite the Tubbs Fire moving at a speed of only one football field per minute, far slower than the Camp Fire, emergency alerts and evacuation orders came too late. Even when they were finally dispatched, the electronic alert systems failed to notify many in harm's way, turning the evacuations into chaos.

Although a fifteen-month Cal Fire probe into the Tubbs Fire concluded that "a private electrical system" was to blame for the fire, not PG&E, the utility company has repeatedly found itself in the crosshairs of what some think is a vegetation problem. With tens of thousands of miles of power lines surrounded by ever-changing foliage, PG&E learned that debris falling against its wires—or sparks falling on the vegetation below—cause most of the region's fires. Even worse, the Camp Fire was not a fluke—such failure of power lines and other electrical equipment regularly ranks among California's top three singular sources of wildfires. Since 2013, power lines that belong to PG&E have caused over 1,500 wildfires. In 2021, a Douglas fir fell on a power line and blew two fuses. The result was the Dixie Fire, the largest single wildfire in California history, which burned for nearly three months, destroying over 1,200 structures, including 650 family homes. The blaze cost over $630 million to put out and destroyed almost 80 percent of Greenville's close-knit, historic town. Its population of 1,000 people about 150 miles north of Sacramento was left in rubble. Some fires have become so large that authorities have grouped them into "complexes." The August Complex was a cluster of wildfires in Northern California that began on August 16, 2020. Over the next four months, the fires would burn more than 1,032,600 acres, becoming the largest fire in California history. The August Complex gave rise to a new classification—the "gigafire"—a fire that burns at least a million acres of land, one level above a "mega-fire," once thought to be rare, which burns more than 100,000 acres.

The phenomenon of increasingly deadly wildfires is by no means relegated to California. While California's wildfires may have grabbed headlines over the past decade, they are hardly exclusive to the Golden State. Decades of vegetation mismanagement; mass urbanization; a warming, drying climate; and strengthening winds have sparked violent blazes worldwide, from the Amazon rainforest to Australia's eastern coast to southern France. In August 2022, emergency services battled once-unprecedented wildfires for a fourth consecutive year as a severe drought and a winter heat wave hit the Gran Chaco, South America's second-largest forest, which straddles Argentina, Bolivia, Paraguay, and Brazil. Even in a country with centuries of wildfire history, Australia's "Black Summer" from July 2019 to March 2020 exceeded the worst-case-scenario predictions. Over 15,000 separate fires destroyed 3,000 homes, killing nearly 500 people and affecting about three billion animals. The fires blasted smoke so high that it damaged the ozone layer.

A year later, 2021 was the second-worst wildfire season in the European Union since recordkeeping began. According to the European Commission's Joint Research Centre, fires were observed in twenty-two countries in 2021, burning over a million acres and killing at least eighty-six, with many more deaths from smoke inhalation still expected. According to a new study that linked the atmosphere's declining humidity and rising temperatures with our planet's increasingly extreme fires, eight of the most extreme wildfire years have occurred in the last decade.

The fires aren't just a problem for the growing number of people who live in their path. Due to the size and intensity of these mega- and giga-fires, their dangers reach hundreds of miles away from the fire zone—and linger long after the flames have been extinguished—in the form of severe air-quality issues. Wildfires release huge amounts of fine particles, a mixture of gas and tiny bits of hazardous matter, often traveling long distances. These particles can cause respiratory problems and exacerbate asthma, heart disease, and other health-related issues for people living in urban areas. The cost of healthcare for treating these illnesses can be overwhelming for low-income people.

To make matters worse, the low-income residents who struggle the most under the burden of these secondary effects of wildfires are also the people who most often lose their homes to California's fires in the first place. This is because the soaring cost of living in California's cities has left less affluent residents desperate to find cheaper housing. Ultimately, many have found themselves forced to move into areas that were previously thought of as too risky due to their propensity for wildfires.

Then, when wildfires strike, they can force the very people who sought housing security on the fringe of urban regions into homelessness. This is true not only for those whose homes burn but also for many in neighboring communities, which suffer a ripple effect caused by fear of future blazes or the withdrawal of fire insurance providers. Often, displaced people seeking a place to live free of the fear of wildfires return to the coastal cities they'd fled in search of affordable homes, putting extra strain on the fragile housing supply. The explosive growth of homelessness in larger cities can be traced, in part, back to wildfires in nearby small towns.

Ultimately, the pipeline from living in a community disrupted by wildfire to homelessness leads to further problems such as mental health issues, crime, sexual abuse, and addiction. Professor Emily Schlickman of the University of California Davis identifies these impacts through research into the effects of wildfires on Butte County's housing supply, focusing on Chico, where homelessness skyrocketed following the Camp Fire. She suggests utilizing federal disaster funds to support voluntary relocation, including buyouts for high-risk homes, constructing affordable housing in urban areas sheltered from fires, and providing insurance and tax incentives for building in safer areas. She reluctantly admits that, in certain instances, entire communities may need to be relocated altogether.

Due to the complex relationships between housing and forest fires, California's ongoing battle with fire is not, as some suggest, a simple matter of climate and geography. It's, in fact, a vexed issue of urban planning, social inequalities, and public safety.

Perhaps most alarming, the wildfires putting cities at risk—whether through direct threat from the flames, as in Paradise, or the larger social fallout to marginalized populations—are getting worse yearly. Each year, tens of thousands of impacted people wonder how these disasters got so bad. Given that Californians and Amazonians have lived with them for centuries, why was it only in recent years that they became so destructive, so frequent, so widespread, and so difficult to escape alive?

Seeing firsthand the desolation that lingered in Paradise, I was determined to understand what was behind these mysteries—and to ultimately answer the question of how we could fashion a fire-resilient habitat.

Zeke, the fire analyst, offered my first clues. As he explained when we met for breakfast in the fire-prone Chico Valley, climate change was only one part of the story: while the added heat, swifter winds, and drier air magnified our wildfires, it was our reckless mismanagement of vegetation that facilitated massive blazes in the first place.

"As long as we try to keep the fire off our landscapes," he tells me, "it will continue its siege. Until *every* last forest town is *burned to ash*." He takes off his newsie cap and sets it beside him.

We were sitting across from each other at a table in a diner a dozen miles west of Paradise—and that town, just out of view behind Butte Creek Canyon, seemed to hang over his every word. As I got to know him better, I learned that Zeke's blunt style was adaptive, the result of a career of ringing alarms to which others too often paid no heed. He'd worked in wildfire mapping and prescribed burning for nearly thirty seasons. At that time, each California fire season had kicked off earlier and lasted longer until it became a year-round phenomenon. The apparent futility of fighting this "siege" had persuaded dozens of his colleagues to leave the field. For Zeke, however, it had only meant that he needed to dig his heels in deeper.

Climate change, Zeke showed me, had undoubtedly made our problems worse. But well before the atmosphere had begun to sizzle, we'd backed ourselves into a corner through misguided fire suppression policy. The Native Americans who inhabited the West understood what forest managers for decades didn't: fires were bound to happen. And rather

than trying to do the impossible—to stop a natural phenomenon—
we needed to take a page out of their book and learn to manage fires
risk.

"The real way to save lives," Zeke says, "is to negotiate a new relationship with fire. A hundred years of waging war on fire, criminalizing its
use, and wiping Native peoples' knowledge off the map have left us with
broken forests and communities on the brink. Sure, we won many battles. But with climate change as an ally, there is no question that fire will
win the war." He studies the artichoke scramble in front of him skepti-
cally, like he has lost his appetite.

The Indigenous tribes that first inhabited the Pacific Northwest all
practiced some kind of controlled "cultural burning," I learn, used to
clear areas of crowded trees, undergrowth, and pests. Fires had always
sprung up across the West, and these burnings made the natural phenomenon more manageable, preventing the dead vegetation that fueled
them from building to dangerous levels.

When European and American settlers first encountered the meadows and clearings of the Pacific Northwest, they saw a bounty of plant
and animal life richer than any land they had seen before. They believed
it to be a perfect representation of an unspoiled, permanent landscape
rather than a delicate equilibrium in everlasting flux. The new settlers
abandoned the cultural burning practices that had protected the region
for millennia. Fire suppression became the favored means of management, which brought dense layers of fallen leaves back to the woods.
Large-scale logging temporarily aided the problem, thinning the forest
in its own way. However, that ended in 1990 when strict environmental
regulations focused on protecting the Northern spotted owl halted the
logging industry overnight.

Since then, the U.S. Forest Service has learned from its mistakes.
It aimed to use scientifically backed sustainable forest management
techniques—including logging, prescribed burns, and thinning—to
treat forest fuel loads and protect vulnerable species like the owl, which
ironically relies on fires for survival. But the prescribed burn campaigns

have struggled to capture the community support they need to be scaled up. Activists who do not want fires anywhere near their property have continued to thwart the Forest Service's efforts.

For decades, there was little downside to the fire-suppression approach. Improvements in technology like higher-flow water systems, bigger fire engines, and planes dropping water from the sky kept in check the fires that broke out. Then, in 2008, everything changed.

Zeke said it was a disaster one hundred years in the making.

Gus Boston, the Butte County battalion chief of California's Vegetation Management Program, walked me through the season that forced a reckoning. He'd grown up in Chico—on the front lines of the fight to suppress wildfires—spending his high school years in Modesto. He was unsure what he wanted to do for a living, so a day after he turned eighteen, his mom sent him off to join the California Conservation Corps (CCC). The CCC's program involved a year of working outdoors to improve California's natural resources, operating under the self-proclaimed motto: "Hard work, low pay, miserable conditions, and more!" After fighting one fire, Gus had found his "more." The searing, smoke-filled air that terrified almost everyone gave him an adrenaline rush like he had never experienced. For the next twenty years, Gus worked as a seasonal firefighter and later as a fire captain, marrying and moving to Paradise, where he and his wife raised two boys. Gus would fight two- to three-acre fires, have the winters off to recoup with his family, and then return to the front lines.

On June 20, 2008, a severe thunderstorm system moved through Northern and Central California, resulting in over six thousand total lightning strikes across twenty-six counties. Along with record dry conditions, the lightning sparked over two thousand fires. Never in the history of the wildfire-prone state had so many fires started in such a short time frame. "It changed me," Gus later recalled to me.

After that, he could no longer take breaks in Paradise. Now, he was an

ER doctor thrust into an unceasing overtime shift. "There was no more fire season," he reflected. "There was only fire."

Having struggled immensely to find a time to meet Gus, his tale of 2008 explained why he hadn't returned my emails for weeks (he quite literally had fires to fight). Why, when he finally did write back, it was in clipped sentences, asking to meet me by the side of a highway (he had little care for modern luxuries). Why he arrived at the agreed-upon mile marker in a Chevrolet Suburban rigged atop enormous lug-lined wheels, whose engine was so loud that I heard the vehicle coming before I could see it (he was moving from one off-road job to the next). I don't normally meet men I don't know on the side of remote highways, but I was glad that I made this exception.

We set off in his truck toward one of his most recent prescribed burns at Loafer Creek, one of many dense forests in Butte County, thick with young trees and brambles, where Gus is cutting and burning fuels to reduce the territory's fire hazard. We walk through a new clearing in the forest, our feet crunching the burnt brush. Only mature trees remain; their bark blackened up the first three feet of their trunks. Gus is proud of his work as the "burn boss" of this site, but he's also quick to put it in context, showing why his efforts aren't nearly enough. "Sixty years ago, we had fifteen to twenty stems per acre," he says, referring to trees that can turbo-charge wildfires. "Now our research shows we have nearly seventeen hundred per acre."

To be effective, these burns need to proceed through three stages. The first burns of an area merely remove vertical vegetation—the canopy. What's left of it will fall to the forest floor, where the increased sunlight allows new plants to grow horizontally, creating the ideal conditions for an unstoppable forest fire. This, Gus explains, is why the region's rampant wildfires over the past decade have done so little to mitigate others that followed in their wake. It's only through the iterative rounds of burning that vegetation becomes controllable, producing less smoke and less damage.

The work is tough, the labor often underappreciated and underpaid, and the scale overwhelming. Yet what I realize while shadowing Gus is that public relations is the most arduous of all. To protect people and property, he and his colleagues must educate their neighbors, somehow persuading people who have been deeply traumatized by deadly wildfires that their best means of protection is through prescribed burns. Even worse, they must come to the edge of their neighborhoods for the burns to be effective.

Leaving the truck behind at the side of a dirt road, we hike into the ashy brush. "This way," he tells me, pulling an adolescent tree to the side like a curtain to make a narrow space to jump through a dense grove.

The Camp Fire sent things over the edge for Gus. After the blaze tore through Paradise, burning down his sons' school, his home was the only house left on his street. One reason was that he had extensively cleared vegetation around it years before. The other was sheer luck. But looking out of his kitchen window to see excavators tear down the rubble of his neighbors' homes, the feeling felt like something else—guilt or perhaps regret.

In the immediate aftermath of the Camp Fire, Gus and his wife sent their teenage sons to live with relatives out of state. "I didn't see them," Gus says between breaths as we scale a rocky hill made slippery with ash. "For months." He, of course, needed to fight the fire. He did it knowing his children were pushing through the hardest times of their lives, now without school, sports, or friends.

Gus's sons never returned to Paradise. One finished high school and started college out of state, the other moved to Europe. And while Gus is proud that they managed to find lives for themselves elsewhere, he's angry that the Camp Fire stole their future. When he struggles to sleep, he thinks about his "happy place," a small cottage beside a river in northern Oregon, with a little fishing boat and fresh mountain air, a place his sons can visit without reexperiencing the events of November 8, 2018. Sometimes, though, he sees darkness climbing behind the mountains, a growing column of smoke.

Later, he tells me about what he experienced as the fire subsumed his town. The screams for help on his radio channel for families trapped by flames he knows couldn't be saved. Skeletons pulled from burned-out bulldozers trying to hold the line—children's faces when they were told what happened to their parents.

Butte County, even before the Camp Fire, struggled with disproportionately high levels of childhood trauma. In 2013, an alarming 76.5 percent of its residents—nearly 100,000 people—reported experiencing one or more Adverse Childhood Experiences (ACEs) before reaching adulthood—the highest ever reported in the state. In the wake of the fire, PTSD advanced to levels comparable to those experienced by veterans returning from war; depression loomed over many who had lost their homes, pets, and communities in a single sleepless night.

While many survivors recovered, the day lives on in memories triggered by the smell of smoke or the sight of a demolition project. The trauma is a challenge that Gus and many of his firefighting colleagues wrestle with every day, and it's also part of the reason he has so much trouble persuading his neighbors of the wisdom for prescribed burns.

After an hour of tripping over piles of unsteady char, Gus tells me we're close to what he wanted to show me. He gestures toward a ridge, and I pick up my pace, my ashy hiking boots squeaking across the charcoal. From the overlook, I see an unending expanse of red soil and massive sycamores spaced dozens of feet apart. Wide shafts of light reach through a mosaic of patches in the canopy toward sprouting wildflowers. For the first time in the day, I hear birds.

This was one of Gus's prescribed burn sites a few months prior. The third fire he'd set on the land that year had crawled across the forest, burning the last layer of thick, yellow brush and revealing an inch of nutrient-rich soil.

Gus takes a drawn-out breath, letting it out in a slow crescendo. "This," he says, "is what the landscape looked like over a hundred years ago."

I sit on the soft soil and close my eyes, listening to the birds chirp. A breeze whistles between the trees. "Will we ever get here?" I ask. "Across the region, I mean."

Gus shakes his head, bending to the ground and kneeling beside me. "We have to," he adds.

Gus's prescribed burns have proven effective, saving hundreds of acres of the Plumas National Forest when the Bear Fire of 2020 surged to the edge of one of his burn sites from the previous year, stopping it in its tracks. But because of shortages in trained firefighters, limited budgets, and the public's resistance, they have proven grossly inadequate. Between 2017 and 2020, Cal Fire and the U.S. Forest Service completed or assisted with prescribed burns on approximately 80,000 acres annually, well short of the agreement these agencies had reached to treat one million acres per year by 2025. In 2020, they recommitted to the one-million-acre target admitting they'd likely only be able to treat up to 400,000 acres annually. And as ambitious as the million-acre-target is, experts warn that it's not nearly enough. Estimates suggest that tens of millions of acres need to be burned statewide. Without consistent funding, trained personnel, and political will, even Cal Fire's minimal targets remain forever out of reach.

In Silicon Valley—a region Gus informed me was very much at risk of burning—I met someone who believed he might have the answer. Anukool Lakhina had learned that the wildfires were something he could no longer ignore during the same event that ripped Gus's family from its roots, the 2018 Camp Fire. While Gus pulled an all-nighter trying in vain to establish yet another line that could stop the blaze, Anukool found himself racing to locate air purifiers to keep his six-year-old daughter safe. Anukool was nearly 150 miles southwest of Gus, but the smoke had overtaken both of their homes. For almost two weeks straight, dangerously high levels of fine particulate matter blanked much of the Bay Area. The pollution was normal in Anukool's hometown outside of

New Delhi, but it wasn't something he was used to dealing with in his new home of San José.

As Anukool reflected on the week's events, one thing was certain—he would no longer ignore wildfire risks. It quickly became a topic of dinner table discussion. His wife, Shefali, enrolled in an Australian Ph.D. program to study what happens when refugees displaced by disaster re-encounter disaster in their new homes, not knowing that her field of study would soon consume her lived experience. Seeing how the smoke had affected his daughter, Anukool decided that he would do everything in his power to find a way to stop the fires from affecting anyone else.

In September 2020, the smoke returned to San José. As almost 500,000 acres blazed across the Pacific Northwest, an unprecedented storm of floating ash streamed across Silicon Valley on its way to the San Francisco Bay, converging with the city's iconic fog, casting an eerie dark-orange glow. Later, a research paper published in the journal *Environmental Pollution* found that California's wildfires in 2020 caused two times as much greenhouse gas emissions as the state had reduced through a tapestry of environmental programs between 2003 and 2019. In other words, the 2020 wildfire season wiped out sixteen years of California's progress in reducing emissions.

Anukool and Shefali founded Wonder Labs, a philanthropic organization funded by the 2017 sale of Anukool's company, Guavus, a $215 million pioneer in real-time big data processing and analytics. The new venture was dedicated to co-developing social, ecological, and technological innovations to mitigate wildfires and related hazards. Thrusting himself onto the front lines, Anukool quickly learned about the potential—and the challenges—of prescribed burning. Even when agencies had sufficient funding, expertise, permits, and support from the public, he found that the constraints of short weather windows shackled them. If forest managers conducted burns when the landscape was too dry, they caused irreplaceable damage to the valuable topsoil layer. The fire wouldn't burn enough vegetation if the lands were too wet. Too windy, and the fire could escape its set bounds, inadvertently

creating the uncontrollable, violent blaze that it was supposed to mitigate in the first place.

As Anukool and Shefali developed Wonder Labs' first product, they
learned everything they could about Indigenous cultural burning. This
practice was criminalized long before the state came to recognize its
value. Anukool studied the techniques of California's Prescribed Burn
Associations. This community-based, mutual aid network helps private landowners put "good fire" back on the land, and also that of the
Hoopa tribe, whose members taught Anukool about their 13,000-year
history of using small burns to renew local food, medicinal, and cultural
resources; create habitat for animals; and reduce the risk of larger, more
dangerous wildfires. He realized what many Hoopa members have understood for decades: that as Western science awoke to the Indigenous
knowledge that it had marginalized for so long, it faced monumental
challenges that tribal leaders could have helped it avoid—millions of
homes, tens of thousands of miles of infrastructure, and numerous factories sprawled across California's mountainous and flammable wild-
lands. Although California's disaster planners had much to learn from
the Indigenous practices they had for so long spurned, it was far too late
to simply copy and paste the approaches that modern development had
overrun.

So Anukool and Shefali went back to the drawing board. Unlike the
twentieth-century engineers who had sought to bend the landscape to
their will, they envisioned technologies that could support its natural
cycle of burning and restoration. And to do so, they realized they needed
to leverage the insights they learned from California's Prescribed Burn
Associations and, later, the Hoopa's and other tribes' success managing
the fire-prone land surrounding them. The goal was a safe, clean, cheap,
mechanized prescribed fire that you could operate at scale. In addition,
it would need to overcome a new challenge with which Native tribes
hadn't needed to wrestle: whereas the Hoopa people understood the
paths of wildfires, avoiding settling in the most vulnerable patches of
forest, modern development was recklessly haphazard, putting towns

like Paradise in the hearts of valleys that frequently burned. Anukool
and Shefali realized that the only technologies that could truly protect
the millions of homes scattered across the West would need to operate
up to the edge of human settlements and even within them.

It seemed an impossible dream until they brought in a pyro-physicist
partner, Dr. Waleed Haddad, who had been drawing similar conclusions
in his work. On top of the trauma-fueled resistance to prescribed burns
near properties that had so often derailed Gus's efforts, burns within
settled communities posed obvious risks. After a series of failed experiments, Anukool and Shefali almost gave up. Perhaps the challenge was
too great—no team of engineers, no matter their commitment, could
save California from its centuries of thoughtless design.

Eventually, Waleed, who Anukool called "the brains," had an idea.
Instead of looking to manage vegetation through massive blazes, they
could build a tool for smaller, more manageable ones. Even if the work
of an individual machine would be slow, it could be mass-produced, creating scale in the form of an army of burn managers.

They call the product that resulted from this thought experiment
BurnBot. It works like a Zamboni, but instead of resurfacing a small
amount of ice underneath the machine, it safely burns the enclosed region that it crawls over. The unwanted fuels are ignited using an array
of high-temperature torches while the flames and embers are fully contained within its "fire box" to ensure no risk of escaped fire. BurnBot
also has a smoke capture and filtration system—something Anukool was
adamant about—to mitigate the release of smoke into the atmosphere.
When the rear of the BurnBot passes over an area, it extinguishes any
remaining embers, further ensuring safety.

Given how rugged much of the vulnerable terrain is in the Sierra
Nevada mountain range, any mechanized system would need to handle
rugged topography and steep grades and break down thick vegetation
while remaining upright. The result is a remote-operated tractor that
can carve perfectly controlled burn lines across the landscape or burn an
entire area in an efficient checkerboard pattern. "No longer will forest

managers need to rely on landscape features as control lines, like rivers or roadways, which can limit their work," Anukool tells me, showing me how you can program the machine's GPS. "We can set this guy loose in your backyard and remove all the dangerous brush without risking damage to your apple tree."

BurnBot aims to overcome four core challenges with traditional approaches to prescribed fire. First, it's far less labor intensive. Whereas it would take years to train a workforce of prescribed burners large enough to get the West's wildfire problem under control, Anukool plans to manufacture enough machines to put a dent in it quickly. Given that current approaches treat only about one percent of the acreage that needs attention annually, hiring our way out of our wildfire problem would be impossible. Second, BurnBot could free land managers from the limited weather windows that now curtail them. "Ironically," Anukool explains, "the best time to burn is actually in the summer when fuels can be consumed quickly, but that is also when wildfires surge when burn bans tend to be put into place." As the fire "season" becomes more and more of a misnomer, growing with each passing year, the periods left for prescribed burning continue to shrink. BurnBot unlocks the potential of all-weather burning.

The technology also could address the most immediate problem with typical burns—the risk of escape. Due to smart planning, very few prescribed burns—less than one percent—escape to become wildfires. But the costs of the few that do are enormous, and the risk and liability of an escape prevent communities that desperately need better fire control from pursuing the approach. The state now provides burn bosses up to $20 million in liability coverage in California. However, other fire-prone states like Oregon, Utah, and Montana don't have the tax base to support that provision. As a result, most burn bosses are one escaped fire away from losing their entire career, threatening this critical workforce's future.

Finally, BurnBot sidesteps the dangers that smoke poses to—and the PTSD-related fears it inspires in—surrounding communities. Smoke from prescribed burns can send wildfire survivors into a panic, ruin the harvests for an entire region of farmers, and, as Anukool learned, put people in the emergency room. Data shows that the smoke caused by wildfires and prescribed burns is often more dangerous than the flames: of the nearly five hundred people who died in Australia's "Black Summer," more than 90 percent perished from smoke inhalation. And this smoke endangers even those who escape its worst effects. In a recent study, a researcher at the Center for Studies of Air Quality and Climate Change found that "smoke samples taken from the air more than five hours *after* they were released from a fire were *twice* as toxic than when they were first released."

Such was the case in Canada's 2023 wildfire season, when the most land burned in any year in the nation's history sent an immense toxic cloud across the world. The area burned even surpassed the 2020 U.S. wildfire season, making it the largest recorded wildfire event in North American history. Over 40 million acres burned, an area roughly the size of Florida.

The fires forced almost 200,000 Canadians to leave their homes, and tragically, at least four wildland firefighters lost their lives in the line of duty. Even beyond Canada's borders, thick smoke engulfed people's homes—stretching from the northern regions of the U.S. to New York and even as far south as Georgia. Air quality alerts prompted concern for the well-being of an estimated 120 million Americans, more than a third of the U.S. population. As the plume of smoke wafted across the country like a toxic belt, residents found themselves forced to breathe smog that was classified as "unhealthy" according to the Air Quality Index. New York City, shrouded in dense smoke, experienced its most extreme air pollution in the city's history, casting an otherworldly orange hue across the sky. IQAir, a Swiss air quality technology company, revealed that out of cities across 100 monitored countries, New York City ranked second only to New Delhi for having the most severe air pollution.

Wildfire smoke is already proving to be up to ten times more harmful to people's health than similar pollution levels from other sources, like factories and car exhaust. Although there is no consensus as to why wildfire smoke is more harmful, scientists hypothesize that it's because of its "finer" particulate matter (PM2.5), which is thirty times smaller than the diameter of a human hair follicle. These particles are small enough to penetrate deep into the lungs and pass directly into the bloodstream, jeopardizing other organs. Unlike prescribed burns, which use noxious accelerants such as diesel fuel, inevitably contributing to the growing smoke pollution in the atmosphere, the BurnBot's high burn temperature creates less smoke, and its filters further mitigate smoke risks. For their first test, Anukool and Waleed put BurnBot to the ultimate test in Anukool's downtown San José backyard, carefully burning the entirety of it. Not a single neighbor noticed the pyromaniacs at work.

The scope of the world's wildfire problem, in which the higher temperatures and swifter winds of climate change only continue to escalate, reaches beyond the grasp of any single solution. BurnBot's scalable precision makes it an essential tool for managing fire fuel within and around settlements. But despite Anukool's ambitions, it may never prove feasible to manufacture enough machines to crawl across California's millions of square miles of wildfire-risk areas. Where large natural barriers like barren ridges and wide rivers allow for it, prescribed burns, completed before the wildfire season, will offer the most cost-effective solution for vast expanses of undeveloped land. Just as it takes a range of tools to fight a fire when it sweeps across a town—from hydrants and hoses aimed at strategic points to fire engines prewetting areas out of reach of the pipe—so, too, will it take a range of methods to prevent wildfires in the diverse habitats that we call home.

Achieving proper safety from wildfires also requires more than preemptive measures to mitigate fires from spreading. We must learn from the mistakes that developers of earlier decades ignored when they built new towns with only one road in and out, instead confronting the unimaginable in the early stages of city planning and crafting plans for how to reckon with it. Rather than believing that we can be immune from wildfire, we should chase the impossible—immunize ourselves through technologies like BurnBot and practices like the Hoopas—while also preparing ourselves for the reality that we will sometimes fail.

With this in mind, we need a series of fail-safes: approaches like Gus's and Anukool's for reducing the damage of wildfires and tools for catching the catalysts that spark fires in the first place—and, when each of those tools falls short, others to help people escape the blazes alive.

Satellite-driven technology identifying potential fires before igniting is vital for heading off the most dangerous blazes. An Amsterdam-based software company has developed the most promising example, Overstory, which analyzes imagery of vegetation to identify the areas most likely to spark. Indra den Bakker, the company's founder and CEO, is a deep-learning engineer who started his career by helping companies measure deforestation, using satellites to show how many acres of forest they were losing in the hopes that they could address its sources.

Indra had always found beauty in the intricate web of vegetation, which twisted and turned in a never-ending cycle of growth and decay. With the help of satellite imagery, he scrutinized this natural phenomenon tirelessly, seeking to unravel the secrets hidden within its densely packed foliage and tangled crowns. However, one day his research took a dark turn. He stumbled upon a disturbing truth: the very vegetation that he had come to love was causing an alarming rise in wildfires, due to conflicts with power lines—a trend that was proving difficult to halt.

It was different in the Netherlands. In 2009, when Indra was a teenager, he witnessed a world first: the Dutch national electricity grid operator had declared it would place several 380,000-volt electric cables underground. The first time such high-voltage cables would be laid underground throughout a city, the project would not be cheap. Indra remembered that the underground wires cost €12.3 million ($13 million) per kilometer, some six times more than overhead cables. But city officials maintained that the experiment was worth it to save highly preventable deaths from downed or faulty power wires.

Disturbed by the growing wildfires he observed through his satellites, Indra changed his focus. It was the first time he recognized the contentious relationship between trees and wildfire. Trees in the wrong place—such as within striking distance of power lines, where a fallen branch can catch fire, or along evacuation routes, where they can block fire engines and strand those attempting to flee—pose great risks regarding wildfires. On the other hand, the right trees in the right place can hold back fire, keeping out the ground cover that fuels blazes. Growing up in the Netherlands, Indra saw the advantages of burying power lines, but his early exposure to their price also showed him why it wasn't always a good solution. They proved extremely costly and often damaging to the environment, especially in protected areas. So he was not surprised to see, as he launched his career measuring deforestation, that Dutch utilities operating in rural locales had begun to turn to extensive vegetation management to prevent tree–power line conflicts. Studies have borne out that clearing brush and pruning trees in the 200-foot "ignition zone" around homes can save a large number of structures from fire while avoiding the extensive environmental damage of digging because they don't sever the "wood wide web," an underground network of fungi in which plants trade resources through root systems.

The stakes of these tree–power line conflicts came into stark relief months after the tragic Camp Fire. As families mourned the hundreds they had lost and survivors like Zeke and his family sought to build new lives, investigators scoured the charcoal landscape for clues about what had unleashed the disaster. There was little doubt about the cause: a hundred years of suppression-driven wildfire policy that had replaced what might have been a minor fire with a bomb of nuclear proportions. But the catalyst itself—what tipped the first domino—was uncertain. Six months later, Cal Fire found it. Tucked into the craggy fir and pine woods of Plumas National Forest, a failed component on an electrical transmission tower slung toward the dry brush below, throwing off the sparks that would forever change California. What could have easily been prevented by the thorough clearing of vegetation had forced more than fifty thousand people to flee and taken nearly a hundred lives.

Why had PG&E, the electric utility that owned the structure, missed this grave risk—and how could it prevent its infrastructure from sparking the next deadly wildfire? Given the trees' growth rate, Indra learned that existing technology wasn't up to the task. It was near impossible for the company to spot vegetation-related risks to its power lines before it was too late. With millions of miles of power lines crisscrossing ecosystems, it takes companies years to scan their entire grid. By the time they've collected the information they need to prioritize maintenance, the recorded data is years outdated.

Given that we remain decades from decarbonizing the energy sector—and some cities can never be fully decarbonized—this infrastructure is here to stay. And, as research into renewables shows, we must learn to live with it. When cities worldwide achieve their renewable energy targets, evidence suggests they will need to run the energy through even more power lines to transport it from where we harvest it to the cities that need it. A Princeton University study found that for the U.S. to meet its net-zero targets by 2050, it would need to increase its high-voltage transmission lines by roughly 60 percent.

Recent disasters drive home why we cannot simply switch to renewable energy machines without also reimagining our power grid. As renewables increase our reliance on power lines, utility companies too often overlook the vital task of clearing vegetation from their rights-of-way. Utility providers argue that they've been put in a predicament, required to ensure the safety of their expanding power grids while complying with costly government directives to adopt renewable energy sources. When they fail to strike the right balance, they put communities in danger.

The predicament became all too real on August 6, 2023, when a series of wind-driven wildfires erupted on the western coast of Maui,

Hawaii. The devastating fires led to evacuations, widespread damage, and the complete destruction of the beloved historic town of Lahaina. More than 2,200 structures were reduced to smoldering ruins, resulting in a staggering estimated loss of $5.5 billion. Lahaina, the vibrant town which harbored a thriving community of 12,000 people, faced utter annihilation. According to data provided by the FBI, the fires claimed the lives of at least 115 people. Surpassing the devastation caused by Camp Fire in 2019, which destroyed Paradise, the Maui wildfire now holds the grim title of the deadliest in modern American history.

Exactly what sparked the fires is still under investigation, but within three weeks of consuming Lahaina, Hawaiian Electric Company (HECO), the utility responsible for supplying power to 95 percent of Hawaii, found itself in the midst of legal battles. Lawsuits flooded in from affected residents, property owners, and business owners, as well as from the county itself. Though the flames have been extinguished, the fight for justice continues.

The suits argue that not only HECO, but also notable landowners such as the state and county, bear responsibility for the devastating consequences of the fires. They allege that negligent mishandling of electrical equipment, failure to regularly remove dry vegetation, and a lack of de-energizing said equipment during hurricane-force winds all played a role in sparking the fires. Maui County, in addition, accuses HECO of "intentional and malicious" mismanagement of power lines.

HECO acknowledged that contact between its power lines and surrounding vegetation sparked the initial wildfire, but it denies responsibility for the catastrophe that followed, instead pointing a finger at the county's firefighters, who declared the fire under control and departed before it resurged.

Even if HECO manages to prove that it doesn't bear full responsibility for the wildfire, there's little doubt that it saw warning signs. In 2019, in the wake of one of Maui's most devastating fire seasons, HECO recognized the need to address the issue of power lines emitting sparks and other related problems. Inspired by California's efforts to combat wildfires, the utility developed a comprehensive plan to enhance safety measures. This included the installation of insulated conductors, fire-retardant poles, and advanced monitoring technology.

Despite its stated intentions, HECO made minimal progress in pursuing these goals over the next two years, allocating less than $250,000 toward wildfire projects. It wasn't until 2022 that the company sought permission from the state to increase rates in order to fund wildfire prevention. Ironically, the company's attention was largely consumed by a mandatory push to pursue another environmental mandate, ramping up renewable energy. In 2015, Hawaii had become the first in the U.S. to pass a law mandating a fully renewable grid by 2045. Regulators completed a comprehensive restructuring of the regulatory framework, incentivizing HECO with significant bonuses for timely completion of green energy projects and imposed fines for missed deadlines. Yet they failed to provide similar incentives for utilities to overhaul how they managed vegetation.

In a striking parallel to the home of Battalion Chief Gus Boston, which was miraculously saved from the devastating Camp Fire through meticulous vegetation clearance and a stroke of luck, there is one house on Front Street in Lahaina that emerged unscathed. Aptly named the "miracle house," this historic wooden dwelling stands untouched while its neighboring homes have been reduced to ashes. The house's survival can be attributed to its recently installed corrugated metal roof, designed to repel flaming debris, and a carefully maintained perimeter cleared of any vegetation that threatened its safety. As highly flammable invasive grasses fueled the fire in so many other areas of Lahaina, the native, impeccably managed trees and bushes around this house formed a protective barrier. The homeowners now aspire to use their house as a model for rebuilding Lahaina.

But perhaps the miracle house has even more significance. It may hold the secrets for bridging goals that might otherwise appear to compete.

The challenge for Internet of Nature (IoN) innovators attempting to solve the puzzle at the nexus of power and wildfire is a common theme: they need to advance sustainability and safety without rolling the technology back. The question for power technology is how it can balance our electricity needs with our need for vital, oxygen-producing forests.

So deep-learning engineer Indra embarked on what would become Overstory. The product uses the highest resolution satellite imagery in combination with advanced AI breakthroughs that have made analyzing satellite imagery practical in recent years. Indra pinpoints dangerous trees and bushes that require immediate attention by providing information about the vegetation, like their current height, growth rates, condition, vitality, and species. He can also predict risks six months, one year, or even two years in advance, giving utilities the guidance they need to curb them.

Historically, tree trimming followed a basic rotation, with each area cut back at even periods, which ignored the enormous differences that the features of microclimates—like additional sunlight or rain—have on plant growth. As real-time satellite imagery and new machine learning methods penetrate the market, Indra and his peers have unlocked two mutually supportive avenues: optimizing management schedules around evident risks and transforming management from a reactive to a proactive science.

Understanding condition, vitality, and species information are critical to this process, something Indra's technology tackles in a way that is now scalable. Species information allows utilities to assess the particular risk of a tree. By accurately identifying the predominant tree species around a power line, one can effectively predict the future dimensions of the forest. As it turns out, not all tree species are a danger to power lines—some are fire-resistant and not prone to falling, making them a critical asset for keeping out the vegetation that facilitates wildfires.

From Indra's perspective, the utility companies whose failures have killed thousands caught in the smoke of wildfires—and perhaps thousands more who breathed the resulting particles continents away—may be our most powerful allies in cultivating fire-safe forests.

In anticipation of the 2022 wildfire season, PG&E boosted its annual wildfire-mitigation spending to nearly $6 billion, an increase of more than a third from the 2020 budget. The mitigation plan confirmed the vital role that technologies like Overstory would play in reducing wildfires: it allocated funding to underground only 175 miles of power lines in high-fire-risk areas, leaving thousands of miles of electrically charged lines hanging across the forests where they could spark another devastating wildfire. The plan also pointed to a new IoN tool that could identify fires before they became impossible to control—wildfire detection and monitoring cameras that scan the mountaintops, applying advanced machine learning to detect the first signs of smoke and instantly warn fire agencies. Called Pano AI—referring to its panoramic view—the technology will work throughout the night even when traditional fire-tower observers wouldn't be able to see smoke columns, feeding information to people reviewing the AI detections in real-time hundreds of miles away. These people can then filter out false positives and improve the algorithm with each "smoke" detection. PG&E budgeted for ninety-eight of these devices.

The goal, says Sonia Kastner, founder of Pano AI, is to accelerate the adoption of new firefighting techniques that emphasize using heavy artillery for aerial firefighting, like aircraft and helicopters, well before a fire spreads out of control. Only a year after its conception, it proved successful. During a pilot with PG&E in August 2021, a camera in Howell Mountain, one of Napa Valley's most prestigious vineyard hillsides, spotted smoke a minute before the fire dispatch did. Another client, Portland General Electric, notified the U.S. Forest Service of smoke from a different fire 104 minutes faster than any of the utility's other techniques. With the speeds that wildfires spread, these minutes often mean the difference of thousands of acres burning, saving lives across the county. By 2024, PG&E aims to install around six hundred panoramic cameras, covering 90 percent of the high-fire-risk areas it serves.

Sonia says her objective is to detect fires within the first ten to fifteen minutes of them starting. Later that day, I would meet someone with a similar vision. "If you can detect a fire quickly," he would tell me, "I can put it out—*before* it becomes a problem."

For as long as wildfires have threatened those living in the West, lengthy response times have proved deadly. Although firefighters stay up around the clock, ready to deploy at a moment's notice, their tools too often render them ineffective in the face of fast-growing fires. Given the rural nature of the areas in the West most prone to fires, the closest responders frequently find themselves dozens or even hundreds of miles away from detected blazes, meaning they cannot approach them in sufficient numbers for hours. And the fires, driven by swift winds, grow far faster than these responders can mobilize. The Camp Fire, for instance, even in its first hours, grew at the rate of a football field per second. In many ways, the fatal lag time with which firefighters mount a response is embedded in the striking geography of the landscape. The deep canyons of the Sierra Nevada, which produce the winds that spur wildfires, result in maze-like routes of switchbacks and towns of dead-end roads. While firefighting airplanes offer a means of dropping water on the hard-to-reach areas where the fires often start, they suffer from the other delay challenges inherent in aircraft: large crews, high costs, and extensive safety checks before take-off.

The shortcomings of fire response collided as Maxwell Brodie, a twelve-year-old boy in Kelowna, British Columbia, found himself doused in falling ash. As smoke approached, Max and his father came out of their house to find the sky above it so filled with gray dust that it looked like the middle of the night. Max turned on the garden hose and began dousing the house's roof. A police officer pulled onto the street, screaming through the loudspeaker that they needed to evacuate. Max and his father sprinted inside to assemble an emergency bag. As they made their escape, they discovered they were trapped in the middle of the 2003 Okanagan Mountain Park Firestorm, which forced more than thirty thousand people to evacuate. The one-in-a-century inferno that began with a lightning strike had been turbo-charged by ferocious winds and one of the driest summers in a decade.

Ephraim Nowak, a childhood friend of Max who grew up a mile away, had a similar experience. It drove Ephraim to dedicate his career to building technology for public safety, developing innovations like underwater robots for search and rescue and drone navigation systems for combat scenarios in which enemy forces jammed GPS navigation. Eventually, Ephraim developed a new approach for mapping wildfires in real time. When the friends reconnected sixteen years after their scrambling escape, they instantly realized they shared a vision for what technology should accomplish.

By then, Max had lived in Silicon Valley for five years, breathing the smoke from a succession of supposedly "once-in-a-century" wildfires. They each knew all too well the damage these fires wrought. And given how the changing climate was helping the blazes travel faster than ever, they understood the desperate need to build tools that could respond before the fires grew impossible to fight.

The venture that they grew out of this vision, Rain, builds from the core principle that IoN innovations all share: the reality that, no matter how sophisticated the contraptions we design, no matter how perfectly we engineer our habitats, we will never abolish the dangers of the natural world. As a result, we must pursue solutions that mitigate those dangers, that allow us to thrive within their uncertainty.

"Rain," printed in size-72 Times New Roman on a piece of paper taped to the inside of the glass door, tells me I'm in the right place. I knock and catch a flurry of robot prototypes off guard, as they amble around the small square office. Bryan Hatton, Max and Ephraim's third co-founder, opens the door. Chatting as we wait for Max, Bryan and I realize we attended rival high schools some 2,600 miles northeast of us in southwestern Ontario.

Eventually, Max bursts through a small door in the back of the office, waving us toward him. We step through it into a vast hangar filled with aircraft. The hangar, he explains, is part of a more extensive redevelopment of a former 1,560-acre Naval Air Station Base in Alameda. Decommissioned in 1997, it had supported the Navy through World War II, the Korean War, and Vietnam. The hangar's sliding doors are open, revealing a spectacular panorama of the Bay Bridge and San Francisco's skyline. The view of the center of the American metro most threatened by fire, from a perch saturated in aviation history, feels like a fitting backdrop for the aerial future of California's war with wildfire.

Max shows me one of their "small" thousand-pound development aircraft in the company hangar. With its sleek landing skids and glossy white finish, it looks like a toy in a sci-fi movie.

"How is *this* supposed to save us from wildfires?" I ask, probably too bluntly.

"This five-foot-wide drone," Max says, resting his hand across the nose, "carries enough flame retardant to put out a quarter-acre fire." He's wearing dark chinos, a Rain hoodie whose elbows have started to fray, and sneakers that are threadbare on the toes, like he has spent hours crouched underneath his machines, working out the kinks. "It's got thermal computer vision," he adds, "to deploy the retardant in the optimal zones." He points to the small black sphere on the front, where the cameras operate. He tells me it travels up to sixty miles per hour, fast enough to chase down the source of a fire before it gets out of hand.

The hangar is getting noisy, so we take our conversation outside, walking to the Alameda Point Waterfront Park, a newly finished green promenade along the water's edge built on the former Navy bulkhead, another part of the area's redevelopment. The sun is setting over San Francisco, a deep-orange sky tinted with red. Max, seeming to read my mind, says, "Luckily, not smoke this time," turning to me.

Every fire starts small. As we walk to the water's edge, Max notes that the reason so many have become the unstoppable forces we see on the news—and that more and more residents of the West have experienced firsthand—is that no one gets to them in the first minutes and hours when human tools remain powerful enough to stop them. But it doesn't have to be this way. Using quick-detection technologies like Pano AI and firefighting tools like those from Rain that are easily deployed across even the most inaccessible landscapes, we can suppress the ignitions that otherwise become catastrophic fires. Setting these technologies into motion beside age-old burning techniques like Gus's and precision burning devices like Anukool's build buffers around cities, we can bring more and more people out of harm's way, even in the most fire-prone environments.

The best methods for using Rain's helicopters will be determined through several trials with fire agencies and municipalities over the next few years. They could fly preemptively during potentially hazardous wildfire conditions using their thermal sensors to locate and combat flames when they first erupt. Alternatively, the drones could take to the air as soon as flames are detected by the hundreds of Pano AI and other fire-spotting cameras already positioned throughout California. "We're the rapid response prong of the wildfire crisis," Max says.

Max recognizes Rain is only one in a suite of tools, including prescribed burns, innovative policies, and ongoing forest management. Although each of the founders I spoke to believes extreme wildfire can be solved, none of these technologies—whether prescribed fire robots, vegetation-monitoring satellites, AI-powered watch towers, or fire-suppressing drones—are a silver bullet. Max imagines a future where all these technologies and more have combined to give people layers of security that are normally redundant. Even when an anomalous weather event, a technological glitch, or human error break one of the forms of protection, another will be right there, ensuring residents are safe. The existential dread that families now feel on a hot, dry August day when they see sparks fly from the utility lines overhead will be merely a story that we tell to our grandchildren.

The sun has set, and we're back at the front door of Max's office. I comment on the lo-fi paper "Rain" sign taped on the door, something he says he hadn't even noticed. For the first time, I notice the darkness under his eyes, a testament that this isn't a job for Max; it's who he is. "You've sacrificed a lot to build this, eh?" I ask, my inner Canadian coming out in the presence of a fellow Canuck.

"Sure, but they're nothing compared to the sacrifices first responders—the very people we're trying to help—make every single day. Firefighters have seen their friends not come back. Neighbors no longer have neighbors. Entire communities have been wiped off the map." Max inhales sharply, then holds his breath. "Those are the real sacrifices."

Despite the potential that so many innovations show for keeping wildfires in check, the challenge grows as the drumbeat of development continues across California's inland valleys. With a greater share of the state moving into harm's way, even the greatest technologies seamlessly implemented will not prevent fire from ever showing up at our doorsteps. The ultimate fail-safe—one that we would rather not think about—is a necessary part of any sustainable approach to fire management: a warning system that can avert the disaster that befell Paradise.

While in Butte County, Zeke had shown me a powerful image that spoke to this hole in existing protections against wildfires. The Camp Fire had entered Paradise at eight A.M. on November 8, 2018, and within two hours, people had begun abandoning their cars on Skyway, one of the only through roads that leave the town to the north, due to gridlock and encroaching flames. Watching the smoke subsuming them, these desperate families faced an unenviable choice—whether to roll for a few more minutes in traffic that might end up stopped by a fire blocking the road or else hike off-road and escape on foot. As more and more people opted for the latter, leaving abandoned cars across the roadway, there was eventually no choice.

At the same time drivers began abandoning their cars, a satellite flew over Paradise. Its image showed the leading edge of a wildfire, and in front of it, at least ten so-called spot fires, which happen when winds transport sparks and embers and ignite new fires outside the main fire perimeter. Most spot fires were small, meaning they were unlikely to take down trees that threatened to block the road. The data had all the information evacuees needed to make the right decision. Without it, however, they had no way of knowing if they were surrounded by the firefront, soon to subsume the road or a mere spot fire. Panic began to rule as they waited in traffic, trapped in an information deficit.

"If people had had real-time maps," Zeke told me, "if they had accurate, up-to-the-minute information about where the wildfire was . . ." He trailed off.

The realization was so dark that I tried to push it from my mind. But I couldn't. I thought about the smell of the smoke they must have experienced, the feeling of hot, dry wind on the back of their necks.

"Countless lives could have been saved," Zeke finally said.

John Mills, a forty-year-old Silicon Valley entrepreneur turned Sonoma Valley retiree, learned firsthand of the failures of America's fire-warning systems. Five minutes after he'd awoken to the rhythmic, thumping sound of a Huey helicopter rattling his house one night, he heard another. He ran outside in time to glimpse a third, this one carrying a water bucket. He saw smoke pluming from his neighbor's lot. Quivering, he grabbed his garden hose, dove into his pool fully clothed, and stood there for hours, gripping the hose, as a C-130 Hercules water bomber roared overhead, raining fire retardant down on him. A few terrifying hours later, he emerged, shivering, from the pool, pleased to see that his off-the-grid homestead had been saved. He'd moved in only a month before that, eager to escape from the hustle of Silicon Valley, where he'd built and sold start-ups for over fifteen years. He had no idea he'd be moving to one of the many populated areas in California where he wouldn't be warned until it was too late about oncoming wildfires that might kill him.

Less than a year later, the 2020 Walbridge Fire broke out nearby, sparked by a lightning complex. The Walbridge Fire burned hottest in the remote creek canyons, destroying 156 homes and 293 structures as it surged across western Healdsburg and Guerneville toward John's house. Despite the fire engulfing his driveway, John once again received no warning. Miraculously, his home was spared.

As I turn off the main highway onto Sweetwater Springs Road, driving toward John's seemingly magical homestead, I couldn't help but gawk at the landscape—deep, vine-filled valleys to my left and thick, rich Douglas Fir forests to my right, reaching toward the cloudless sky. After nearly six miles, I think I've missed John's driveway, debating whether I should turn around. Then I see his gate. It swings open, revealing a two-mile driveway that crosses a small creek. While the tall grasses and wildflowers along the curving drive make it a place of great beauty, I can see a few scattered charcoal chunks that remind me of the Walbridge Fire that subsumed it just two years earlier.

John welcomes me at the bottom of his steps in a tweed blazer, dark jeans, gray sneakers, and a tan fedora cocked so far to the right that it touches the corner of his clear-rimmed glasses. After exchanging some pleasantries, we find a place to sit in his fittingly elaborate living room, and he tells me his story. The more he shares, the more I realize why he's so forthcoming. For years until the Paradise disaster, he felt that fire-warning systems were the biggest issue in the world that no one wanted to talk about.

After 2019, desperate not to have to bolt into his pool again, John spent the year going to town hall meetings about wildfires and wildland firefighting training, and emergency management courses. In addition, he worked as Pano AI's interim chief technology officer, helping an old friend get her company off the ground. He also became a volunteer firefighter, buying himself a Nomex fire suit and a Dodge pickup truck to assist neighbors and local squads. "I wanted to understand what was going on on the front lines," he says. "This was a problem that no one else was going to solve for me, and I wanted to do what I could to help others in the community as well."

There was, however, one thing that kept bothering John. Twice now, a supposed "once-in-a-lifetime" wildfire nearly burned him and his house to the ground, both times with no warning. There was no denying that the voluntary reverse-911 programs intended to warn of oncoming disasters had limited reach. Evacuation alerts were sent to the wrong phone numbers or obsolete landlines. Frequent system glitches left thousands of people without warning as flames destroyed their homes and killed their neighbors. And John knew that others hadn't been as lucky as he had.

In Paradise, in the wee hours of November 8, many residents did not realize the danger until they saw the flames and smelled the smoke. The Butte County Sheriff's Department decided to use what experts say is an outdated emergency response system called CodeRED, which notifies only the residents who had previously opted-in to receive phone calls, which, it turned out, was only a small fraction of them. Even among that group, up to 60 percent of calls failed to reach them. The first order to leave came as the fire was already on the edge of town.

Fear of causing panic and traffic jams on Skyway, the one main road in and out of Paradise, led officials not to issue an Amber Alert–style message. Ironically, only the panic of frantic and yelling neighbors triggered people to finally evacuate at the last minute. Dr. Thomas Cova, a geography professor at the University of Utah who specializes in emergency management, has since said that it is a myth—not backed by any research—that notifying the public immediately at times of disaster, whether it be a wildfire or a hurricane, backfires by causing panic and hampering emergency efforts.

The same thing happened on Maui, where, in the midst of a blooming inferno, the sirens remained eerily silent on August 8, 2023. Hawaii boasts the world's largest collection of outdoor sirens—over four hundred. Originally meant for military purposes, these sirens have been invaluable in alerting the public to natural disasters, such as wildfires, as stated on the state's official website. Yet Maui's emergency management administrator made the decision not to activate the sirens. There were concerns that the public, mistakenly thinking a tsunami was imminent, might flee toward the mountainside and into the path of the flames. Despite emergency alerts being sent to various forms of media including cellphones, TVs, and radio stations, many residents reported not receiving any alerts, citing poor service or a lack of cable TV. The tragedy that resulted underlies the desperate need for a multitiered approach to emergency alerts.

John was tired of finding himself in the crosshairs of the public failure of the emergency warning system, so he made it his mission to build a better tool. This system would give Californians, and ultimately, the world, real-time, trustworthy information about fire movement and firefighting efforts. The app that he eventually developed, called Watch Duty, stood in stark contrast to the approach of Cal Fire, which only provides updates about once a day, leaving municipalities—and individuals—to fend for themselves in between. The nightly news, John found, was even worse, giving only broad overviews of the wildfires' paths. Social media was the best source of information, but misinformation and a lack of geographic filtering meant that relevant and accurate information was hard to find. To make that intelligence accessible, John looked to amplify the voices of the people who gave him solace when he had spent hours each day checking for updates: the fire reporters, who keep their ears glued to radio scanners. Many of them already had tens or hundreds of thousands of followers—and most importantly, the community's respect, like the scouts who once lived in lookout towers atop mountains. John just needed to give them a platform to weed out all extraneous chatter and push clear alerts to those needing them.

With almost no capital and only the donations of servers from companies like Amazon Web Services, Salesforce, and Microsoft, John went from having an idea to building a team to deployment to growth in a single quarter. Two days after he launched Watch Duty on August 11,

2021, it had over 22,000 active users, mostly people following the fire reporters' existing social media accounts. Within a week, the audience grew to 40,000, and the app had its first test case.

The Cache Fire swept through the southern edge of Clearlake, 65 miles north of Healdsburg, on August 18, 2021. Because of a Watch Duty alert, schools and hospitals were evacuated forty-five minutes before the government's alert went out. John cried, reading the letters of parents of students and families of hospital patients, knowing that, had they waited for the official warning, they likely wouldn't have survived.

In the wake of that success, John got his first taste of the battle that would escalate between him and state fire officials who had learned to dismiss *unofficial*—and potentially *unverified*—information. While their concerns about misinformation on social media had proved valid, these officials did not account for the safeties that John built into his app to validate the data it spread.

The reporters on Watch Duty embark on a grueling, exhaustive process, monitoring dozens of radio channels and web pages, which takes up most of their time during fire season. Reporters like Sekhar Padmanabhan, aka "barkflight," brought to the app years of experience disentangling fact from fiction, having previously covered fires as a newspaper reporter and then fire tweeter. When gathering news for Watch Duty, Sekhar and the eleven other reporters on the team are each assigned to different districts, using Slack to coordinate, pool, and help each other validate the information in real time before posting anything publicly. Slack bots automatically pull in official updates, along with smoke-detection data from Pano AI and other fire-spotting cameras, making the process of synthesizing information even faster. The reporters are very strict about confirming sources, following a rigorous, publicly available code of conduct about what gets published and, more importantly, what doesn't.

The vetting processes and perfect accuracy of Watch Duty's reports thus far have done little to mollify officials. In June 2022, a Cal Fire dignitary told John he'd be smart to leave town. The night before I met him, at a wildfire tech dinner a venture capital investor put on, John tells me another high-ranking official refused to shake his hand. Others have cast dirty looks at him at town hall meetings, even in his local grocery store. Funding, it turns out, is part of why many officials despise Watch Duty. When the app caused schools and hospitals affected by the Cache Fire to evacuate before they received an official order, Cal Fire missed its window to apply for a $1 million federal grant. While John takes no responsibility for that—attributing it to a broken FEMA system—he sympathizes with individual firefighters' and sheriffs' concerns over maintaining public safety control. "Ideally, we'd work with the government, not against them," he explained. But his experience surviving two fires about which official warnings never came—not to mention the Camp Fire's fatal mistakes—leaves John with little doubt that he has the moral high ground. "I've extended multiple olive branches," he says, "and nothing came of it. This is what happens when a government fails the public. And we should never apologize for saving lives."

"Do you ever get scared?" I ask.

"He doesn't, but I do," John's girlfriend, Yana, walks in to introduce herself before starting dinner.

"It's true—" John says. "I've never taken well to being told what to do." He reminisces about his childhood in upstate New York, building, tinkering, and blowing things up in his backyard, much to his mother's chagrin. "But until I wake up in the trunk of a car somewhere—"

"We're not going to let it get that far!" Yana yells from the kitchen. "Right, John?"

"I'm not going to stop," he continues. "When officials get angry, it reminds me that we're doing the right thing. If the citizens didn't love it so much, and the government didn't hate it so much, it would mean I was doing the wrong thing," he pauses. "Watch Duty is the most disruptive, violent thing I've ever put my hands on. You'll have to kill me to stop me. In these situations, minutes count, and seconds count. People should never die because they didn't get a warning. Give people the facts, the truth, and let them decide what to do."

Since January 2024, only two years after the app's launch, Watch Duty has expanded to the entire state, boasting over a million active users and paying fire reporters salaries through user donations to focus on work they previously did for free.

As the company grows, threatening to drag John from the quiet retirement life he once sought in Sonoma Valley back into days teeming with work, the tens of thousands of letters he receives keep him going. One lifelong Sonoma resident, an eighty-two-year-old grandmother, worried that Watch Duty would only trigger her anxiety, but, as she wrote to him, it did the complete opposite. Knowing a little more, she said, feels a whole lot better than knowing a lot less. Others write in to share that since they downloaded Watch Duty, they've slept through the night for the first time in years, knowing that screams and the smell of smoke won't awaken them. Some have told him that after losing family, friends, and livelihoods, unable to bring themselves to rebuild, having watched their home go up in flames, they decided to stay—because of the peace of mind Watch Duty gave them. It gave those who lost everything but their lives a reason to stay and fight, restoring their hope for a future in California.

The warning system also provides key benefits for forest managers like Gus, whose prescribed burns too often provoke panic, one of the biggest hurdles for building the support necessary to manage the state's forests. Despite Cal Fire's best efforts to send out press releases of upcoming burns to TV and radio networks, print newspapers, and social media, every prescribed burn led to an uptick in 911 calls. He observed the same thing at Sonoma's emergency dispatch center, where John volunteered. Since Watch Duty's launch, 911 call volumes have been drastically reduced during prescribed burns.

"I don't want a thank you," John says, "but I'd also prefer not to be told to go fuck myself when I'm in line at the post office."

As we're nearing the end of our conversation, it dawns on me that John forgot to put fire in the name of his app. "Shouldn't it be something like 'Firewatch'?" I ask.

John shakes his head. As it turns out, the Watch Duty team has big plans. First, to expand to other fire-prone states and provinces, starting with the Pacific Northwest. Then, to other natural disasters facing the American west. "Watch Duty started with fire, but it was never only about that. It was always about the unknown threat that lurked just around the corner, ready to kill us. Earthquakes, landslides, volcanic eruptions, tornadoes, blizzards, tsunamis, cyclones, floods, hurricanes."

He nods his head. Floods are probably what's next, he adds. Wildfires directly impact an area's susceptibility to flooding because the intense heat from fires causes the earth to become about as absorbent as a parking lot. Flood risk remains significantly higher up to five years after a wildfire—until vegetation is restored. In an era where some areas burn multiple times within a few years, many constantly fear wildfire *and* flooding.

As we head down the stairs to my car, I pause beside the pool that had been his refuge during his first wildfire experience, studying it. I see the hose coiled in the corner that he had grabbed on to and used to clear the ash from a patch of air when he would come up for breath.

He notices me staring at the pool and seems to read my mind. "Desperate times call for desperate measures," he says, chuckling faintly but stopping a little too suddenly, perhaps grieving the retirement for which he had built this adult playground—or pondering the fate he so narrowly escaped.

CHAPTER 5

Draining the Swamp
in Real Time

The single raindrop never feels responsible for the flood.
—Douglas Adams

ON A SCORCHING SUMMER evening, June 30, 2021, tragedy struck;
a raging fire set the small village of Lytton, British Columbia, ablaze. The
previous day, temperatures had risen to unprecedented heights: 121.3°F,
the hottest measurement ever recorded in Canada's history. The wildfire season peaked much earlier than usual that year, with drought
conditions and a series of punishing heat waves leading to widespread
fires.

A blazing inferno, whipped by furious winds of up to 44 miles per
hour, raged inexorably toward the stricken village. The valiant efforts
of volunteer firefighters were met with a relentless wall of flames. Ex-
plosions from nearby propane tanks and shifting winds added to the
danger, sending scorching embers flying like hail. Within minutes, fire
engulfed Lytton and its inhabitants, forcing most to flee with nothing
more than their lives.

The fire claimed two lives and left 90 percent of the village and the
homes of the 1,750 First Nations residents who live nearby in ruins.

With nearly every house destroyed, the communities were cut off from vital services such as electricity and water. Every feature of Lytton's Main Street—from essential services like the post office, ambulance station, healthcare clinic, police station, village hotel, and city hall—was leveled. Cultural landmarks, such as the town's Chinese History Museum and sixteen hundred irreplaceable artifacts, were wiped out. Residents were left homeless. Lurking in the wreckage was further danger: chemical runoff from firefighting efforts had contaminated the village watershed.

When all the destruction was accounted for, the 2021 wildfire season proved to be British Columbia's third worst on record regarding the area burned.

Less than five months later, tragedy struck again. Starting November 14, 2021, heavy rains doused many parts of British Columbia—Lytton included—in an onslaught unlike anything the region had seen before. Rivers overflowed their banks from top-level flooding while mudslides blocked already damaged roads. As rainfall continued for days, what was left of any buildings became submerged under rapidly rising waters, roads had turned into rivers, cars floated away, crops died, and livestock washed away or drowned. Most notably, the heavy rainfall and ensuing landslides across the only access roads stranded tens of thousands of people, cutting them off from the rest of Canada. So vast was the damage that much of the mangled highways will not be repaired until 2024.

The close link between heat domes, wildfires, and floods has yet to be commonly known. Yet as Lytton demonstrated spectacularly, a heat dome can trigger an unexpected fire that can potentially create hazardous flooding. It's a sobering lesson in nature's cascading forces of destruction: When fires burn the landscape's greenery, they take the invaluable protection that the vegetation provides against heavy, flood-causing rains. Without plants to absorb water and obstruct its currents, it courses unimpeded toward more vulnerable areas below, leaving devastation in its wake. Even the fires themselves exacerbate flooding conditions, as the residue of flames creates hydrophobic soils.

The growing disasters bearing down on our cities are so interlinked that to survive any one of them, we must learn to grapple with the roots of each. As we saw in Chapter 3, heat waves result in drought, laying the groundwork for wildfires, which raise the risk of flood, the world's deadliest natural disaster. Thus, the central challenge for restoring human habitat isn't simply addressing the problem of insufficient water in the landscape. It also requires understanding the other side of the climate change coin: bigger, once-tropical storms carrying too much water, storms that seem to venture further north each year—into Europe and North America.

Floods are, without a doubt, one of the most immediate threats to the future of our cities in a climate-changing world. With intricate infrastructure systems, high population densities, and bustling business centers, cities are particularly vulnerable to the impacts of flooding and extreme weather events. Various sources can incite destructive floods, including intense rainfall, sudden snowmelt, storm surges, or the collapse of dams and levees. Floods disproportionately negatively impact urban populations due to the concentration of impervious surfaces such as concrete and asphalt in cities, which prevent the ground from absorbing water and diverting it to city streets and buildings. Additionally, the high concentration of city buildings creates less green space, which would otherwise serve as a natural buffer against floods. As a result of the intersection of these factors, flooding in urban areas can lead to intense property damage, displacement of residents, and in severe cases, loss of life.

The best strategies for confronting the impact of floods on vulnerable populations require us to understand the human causes of what is too often termed a "natural disaster." With water levels rising and flooding becoming an increasingly vexing problem for communities across the globe, sustainable solutions must account for the mistakes of prior urban development, substituting systems that used artificial tools to manage flood water with new models that harness the power of nature. Many natural features can help mitigate the impacts of flooding, such as forests, wetlands, and even constructed ponds and rain gardens. These features help to slow the movement of water, allowing it to be absorbed more effectively by the ground and vegetation. As we continue to explore new and innovative ways to address the ever-present threat of flooding, it's clear that harnessing the power of nature will be an essential part of the equation.

On July 13, 2021, Corry Martin and everything she and her husband, Jean, had built for the last twenty-three years was in the wrong place. They lived in the Walloon Region of Belgium, and while they were fully aware of the storm overhead, they never realized how dramatically it could change their lives.

"We saw a little bit of water trickle in through the fireplace," Corry told me, "and before I could grab the mop, the water was up to our ankles." From there, it took only minutes for the water to rise. Eventually, it was just below her chin. The dining table floated across the room. Then the TV crashed off the credenza. She and her husband couldn't open the front door to escape because of all the water weighing against it on the other side. Corry climbed onto a windowsill and cried out for help.

Thousands of people die every year in similar scenes—drowning in houses, stranded on rooftops, or trying to swim through the murky tides that take over their hometowns, only to find themselves crushed by a car floating toward them. When the violent, cascading water finally ebbs, corpses can be left suspended in rotting branches, somber reminders of nature's destructive power and unforgiving swiftness.

Corry and Jean were traumatized. But having survived, they spoke with grace and an almost giddiness, as if grateful they were still here. Their hero was a neighbor, who saw Corry waving from the window and swam over without a second thought. She opened the window and pulled them out. In their telling, Corry and Jean were lucky. The storm's floods claimed 243 lives, with the vast majority in Belgium and Germany.

With an already wet start to the month, extreme precipitation on July 13 challenged records and drenched the ground. On July 14, this created a treacherous combination as higher-than-normal runoff levels caused flooding in areas unprepared for these conditions. While the rainfall was exceptional, some experts felt that Belgium and Germany had no excuse not to be better prepared. Unprecedented floods had killed 232 in Europe less than two decades before, leading countries to partner up and create the European Flood Awareness System (EFAS) to provide early warnings that, leaders claimed, would reduce the impact of transnational floods. One hydrologist who set up and advises EFAS expressed shame that even more people would die under its watch, calling the death toll "a monumental failure of the system."

When Corry and Jean had recovered from their ordeal, they did not doubt that they needed to move somewhere with better flood protection. They did not expect, however, that they would find what they were looking for just across the border in the Netherlands. While they continue to experience similar bouts of extreme rainfall, living in the same climate where they had nearly died, they can rest easy in their new neighborhood, knowing that the floods that destroyed them never entirely submerged any Dutch towns. Not a single person in the Netherlands died.

How did the Dutch escape what the minister-president of Germany called "the disaster of the century"? For Corry and Jean, this was the riddle whose answer helped them overcome the effects of their trauma.

As it turns out, the Dutch have been waging war against mighty rivers and a relentless sea for over a millennium. Three of Europe's largest waterways—the Rhine, Meuse, and Scheldt—empty their waters into the country, leaving 60 percent of its residents liable to floods from below or above. It is the Netherlands' low-lying nature, after all, that gave the nation its name. Trapped in a constant struggle for survival at and below sea level, much of this nation has slowly sunk away, resulting in an ever-changing shoreline.

The country's aptitude for water management has not come without extreme hardship. On the night of January 31, 1953, a violent storm, and the powerful storm surge it brought, blanketed the Netherlands and parts of England, Belgium, Denmark, and France, leaving behind catastrophic flooding for miles along these nations' coastlines. Hurricane-force winds over the North Sea had combined with an unusually high tide to send a wall of water toward each coast. An estimated 30,000 animals drowned, 47,300 buildings were damaged, entire villages were swept away, and over 2,000 people perished, more than 90 percent of them in the Netherlands. It remains the worst natural disaster in modern Dutch and British history.

The 1953 floods were a deeply tragic event for the Netherlands. But the fear among survivors that anyone might be next galvanized Dutch leaders into action in a way that no country had experienced before or since. The major project that came out of the nation's soul searching was an extensive system known as the Delta Works, designed to protect against floods and storms along the estuary of the Rhine and the Meuse. The plan took decades of hard work, with many believing it would never be finished. Still, in 1998 this monumental feat was complete: three locks, six dams, and four storm surge barriers formed a massive chain of coastal defenses—the largest flood protection system in the world.

Since then, hydrologists have pointed to the Netherlands as the gold standard for water management, offering a blueprint for flood protection for cities far and wide that now struggle with previously unforeseeable water disasters. Formed from massive walls with arms wider than two football fields, bolstered coastal dunes replenished annually by 12 million cubic meters of sand, and lined with reinforced dikes and enlarged riverbeds, the Delta Works has even been declared one of the "Seven Wonders of the Modern World" by the American Society of Civil Engineers. That the massive chain of flood protection structures uses state-of-the-art sensors and analytics is a given. As night falls and day breaks, digital monitoring reveals hidden issues in structures that would otherwise go unnoticed, allowing experts to intervene before any damage even begins to risk compromising the system's operation.

It was Room for the River, a supplemental project—the largest Dutch hydraulic engineering project since the Delta Works—that saved many lives in July 2021. Constructed over a decade, the Netherlands finished Room for the River in 2015, combining flood protection, master landscaping, and environmental conservation to protect communities around four main rivers: the Rhine, the Meuse, the Waal, and the IJssel. In July 2021, the volume of water that the Meuse accommodated was both spectacular and dramatic, the ultimate test for a civil engineering project that had polarized a generation of Dutch taxpayers, of which many believed it to be wasteful. As many German towns suffered tragically, the catastrophic storm proved that the widening of the river—combined with natural development along its banks—would work almost exactly as engineers had predicted. Along the approximately 40 kilometers of the Grensmaas, where the Dutch and Flemish banks have been lowered by excavation to address the river's most critical flood point, the water level remained one to two meters lower than it had been during the last massive flood before the project, some thirty years before. It wasn't perfect everywhere—a few places along the river spilled over, showing a need for further excavation. However, decades of deepening and widening the river had paid off. And when the Grensmaas subsided in the days afterward, not only were the millions of residents living near the river relieved that their homes remained intact, but they were also thrilled that the project's extensive new hiking areas along the river's edge could be enjoyed once again on less rainy days.

Out of the crucible of the Netherlands' disaster scenario in 1953 grew insights that are now increasingly vital worldwide. The trends of modern life have made the Dutch approach to flooding unavoidable, as the urbanizing population around the globe continues to relocate tens of millions per year from relatively flood-resistant farmland to ports on the sea and areas where rivers meet. These sites where humans have built cities for recorded history are among the most vulnerable to water, some of the first areas to flood. And as we build them up, multiplying impervious concrete surfaces, we increase the flooding risks even more.

To avoid the increasingly deadly disasters that took the lives of some of Corry and Jean's neighbors, we must prepare ourselves to hold back the new normal of floods. To do so requires proofing streets and buildings for the more extreme weather that greenhouse gas emissions have already baked into our future.

Undaunted by the grim challenge, a range of IoN innovators are confronting this reality, revealing that adaptation proves an indispensable defense far from surrendering to climate change. The formidable challenge of the heavy, frequent storms ahead can be outmatched by our power to design resilience in our habitats.

A series of wide-ranging catastrophic floods have confirmed that what may have once been considered a Dutch problem is now the world's challenge. In the 2000s, Americans witnessed this phenomenon in historic storms ranging from Hurricane Katrina in New Orleans to Sandy in New York and Harvey in Houston. Climate change plays an important role in this phenomenon, even for noncoastal cities that do not have to fear sea level rise. Since the heating atmosphere holds more water, changing weather patterns boost the size of tropical storms. Twenty-twenty was a record year for flooding in the Asia-Pacific region, with cities in East, Southeast, and South Asia seeing economic losses, loss of life, and widespread displacement due to heavy rains. In China alone, 2.7 million were evacuated, and 63 million were affected. While the true death toll from the floods remains unknown, the fact that many officials were later arrested for deliberately underreporting or concealing it speaks to the magnitude of the disaster. As a result of heavy rains, Nepal has experienced a devastating 4,500 landslides; at one point, a third of

Bangladesh was underwater. In the same year, South Asia saw another 17 million affected by flooding, impacting vast regions of Indonesia, Vietnam, and the Philippines.

Much of the devastation of heavy rainfall in cities is due to the prevalence of an obsolete wastewater processing model called a combined sewer system (CSS). These water processing systems drain sewage and wastewater into a single pipe, funneling the combination to water treatment plants. When introduced in 1855, they were hailed as a much-needed evolution away from the cesspool ditches that ran along streets and spilled over during rainfall, sparking deadly cholera outbreaks and other diseases. But what once saved thousands of lives now puts thousands more in danger because of CSS's unique flood risks.

In the United States alone, 860 cities that more than 40 million residents call home have outdated CSS. They are mostly concentrated in the Northeast and Great Lakes regions, though the problems stretch as far south as West Virginia. Across the Atlantic, CSSs are just as widespread. In the European Union, roughly half of all sewer systems are combined. In the U.K., the proportion is even higher—70 percent.

Across Asia, CSSs are practically the default approach. In Southeast Asia, many of the largest cities—including Bangkok, Thailand; Hai Phong, Vietnam; and Manila, Philippines, home to nearly 15 million people—use combined sewers. In Tokyo, the world's most populous city, 80 percent of residents live in wards that rely on CSSs. China's crowded cities also largely rely on these sewage systems. Guangzhou alone releases an astonishing 470,000 tons daily into the Liuxi River, a model reflected in other Chinese megacities such as Shenzhen and Shanghai.

Today, the CSSs across these and so many other cities no longer function as the miracle water-treatment systems that they were designed to be. Built at a cooler time when fewer people lived in towns and when storms carried less of the water and wind that create storm surges, they failed under the power of today's weather. Often more intense, frequent, localized, and much less predictable, recent storms regularly push our sewer systems to the brink of catastrophic failures. During heavy rain-falls, the stormwater overwhelmed the combined sewer pipes under-neath the streets, leaving no space for the raw sewage from homes and businesses. Frequently, sewage either backs up, destroying homes, or the cities that manage the system unleash untreated polluted sewage and garbage, allowing them to flow into street drains and directly into local waterways, decimating marine habitats, which sometimes take decades to recover. During the biggest storm surges, cities typically experience both.

Humans are not immune to the dangerous effects of these failures, as combined sewage overflows (CSOs) can carry bacteria, fungi, para-sites, and viruses that can cause intestinal, lung, and other infections, the disease outbreaks that the technology once prevented. CSOs can also cause beach closures, kill fish, and contaminate drinking water sources.

According to the U.S. Environmental Protection Agency (EPA), CSOs release around 850 billion gallons of diluted yet untreated sew-age into America's surface waterways daily, enough to fill 3,070 Empire State Buildings. When rains get bad enough, residents are told to stop showering, flushing, or washing their clothes or dishes to avoid further aggravating these overflows. As more extreme and frequent storms ter-rorize cities, their danger is clear. But the huge price tag of new sewer systems leaves many city officials hoping for the best and grappling with the worst multiple times a year.

In Chicago, urban planner David Leopold spotted an opportunity to use smarter analytics to stem flooding—and optimize the budget to make it happen. In 2005, David was studying planning when he stum-bled on the concept of "green infrastructure," an emerging method of using nature's best water-absorption systems as a technology to tackle our stormwater problem. Fifteen years later, having become the director of city solutions at the Chicago-based City Tech Collaborative, he per-suaded America's third-largest city to spend $50 million on green infra-structure interventions. These practices harness the power of nature to mitigate the damage caused by heavy rains or stormwater runoff. They include bioswales, rain gardens, and permeable pavers, each of which relies on natural force to reduce pollution and mitigate the flood risks of urbanization. Bioswales are vegetated ditches that capture and funnel stormwater flowing across a city's impervious streets and sidewalks and treat it using plants and soil that filter pollutants. Rain gardens are smaller, shallow depressions that similarly absorb runoff and filter it into aquifers below the city streets. Permeable pavers are interlocking paving stones that allow water to pass through the road into the soil and the groundwater below, reducing runoff. In addition to reducing the impact of floods, these green infrastructure interventions yield other benefits, slowing down the rate at which water flows through streams and rivers, which allows these ecosystems to naturally filter it, allowing water to reach the roots of urban trees, and green spaces for recreation and relaxation.

After years of struggling to find enough funds to address the grave flooding risks he couldn't avoid finding across the city, David suddenly had a new problem: With so many promising technologies now within reach, how could he maximize their impact?

David has good reason to want to make the most of every penny. Since 2015, flooding due to excess stormwater has resulted in over 182,000 claims of property damage in Chicago, with a total estimated cost of $735 million. As any Chicago resident who experienced a backup in their home can tell you, the stakes are enormous.

"It's exhilarating to see Chicago take these first steps and invest in green infrastructure," David tells me, showing me a flow chart of the different pathways he wants the water to drain in the city. "But we're not taking it on faith that it works." His smile fades. "I must justify why I chose a bioswale over permeable pavers or a rain garden."

He realized that to be successful at preventing floods and increasing the quality of the city's bodies of water, he needed to tailor the infrastructure to the community's specific needs and location. Permeable pavers, for instance, worked best on walkways outside the city center, where residents would typically rely on shovels and snowblowers to clear snow. But on roads and downtown sidewalks, snowplows would often destroy them. Rain gardens worked best in the densest neighborhoods, especially when residents expressed an interest in helping to tend to the plants. Due to their valuable space, large bioswales needed to be targeted to the areas with the greatest risk of flash floods.

To spend the $50 million wisely, David collects data about which types of green infrastructure work best in which situations. By replacing one centralized million-gallon holding tank, his program proposes the installation of thousands of smaller bioretention cells, which both store and treat water. While these assets would have previously proved much harder to track than treatment tanks in a single wastewater facility, David's team of computer scientists made it possible through a system combining weather information with surface and groundwater monitoring. Aboveground, the sensors monitor weather conditions to see the outputs the infrastructure needs to manage. Belowground, they monitor soil moisture and water quality to ensure the cells contain the water as designed.

Though I'd learned as I studied ecological engineering to always account for results and continue to iterate, David's commitment felt almost excessive, like a fixation. Later, he would confess that if he doesn't get this right, he worries there won't be another $50 million. "Not just for Chicago," he would add. As one of the first major municipal investments in green infrastructure in a country that has proved reluctant to pursue the technology, Chicago's program, he fears, could determine whether any American cities opt to follow the city's lead.

David walks me through Chicago's current water treatment system, showing me how the data could change it. Using real-time measurements, the city already chooses to hold back or release water to mitigate immediate flood risk. As it installs new green systems, the data provides a feedback loop, testing design assumptions. David traces a circle in the air. "Ultimately," he says, "it's showing us something we didn't expect: that different places flood in different ways." Turning his attention to building the right systems for each neighborhood, he analyzes why.

It comes not a moment too soon. When the skies above Chicago opened in July 2023, they unleashed a record-setting rainstorm that left an indelible mark on the city and its residents. Over 12,000 reports of basement flooding were filed with the city's nonemergency helpline, 311, as the relentless downpour overwhelmed the city's sewer system and the Chicago Deep Tunnel Project. Initiated in 1975, the Deep Tunnel Project aimed to reduce flooding and improve water quality through extensive infrastructure development. However, despite the construction of over 109 miles of tunnels and multiple reservoirs, the project's effectiveness was called into question after the severe July storms caused widespread flooding, prompting renewed discussions about the city's resilience against extreme weather conditions.

"Planning has always been about imagining a future." He raises his palms as if to catch invisible raindrops. "These new technologies are just helping us chart a faster course to get there." He rotates his hands around each other, gesturing with the enthusiasm of an Italian grandmother.

Cities across the European Union have already demonstrated much more interest in the potential of green infrastructure technology than their American counterparts, investing on an annual basis roughly one thousand times the amount of Chicago's plan. Despite this, the EU projects that this figure needs to double within a decade for its systems to keep up with changing weather. The EU predicts that digital technologies—the IoN—are expected to play an integral role in this infrastructure transformation, improving both efficiency and sustainability while allowing for more cost-effective management. Leaders are looking to an EU-funded project called Digital Water City (DWC) to lead the charge of its digital revolution in sewage management.

The initiative recognizes that, as in Chicago, where neighborhoods have different needs, the best approach to wastewater management will look different for each city. For instance, areas that are being developed for the first time can avoid the growing issues caused by CSSs entirely by building separate sewer systems (SSSs). These ensure that runoff created by high-intensity rainfalls is directed toward nearby water sources without obstruction, bypassing the environmental risks of raw sewage in waterways and the destructive backups that haunt cities using CSSs if managers don't release the raw sewage quickly enough. Unfortunately, replacing CSSs with SSSs isn't feasible in many European cities. In these densely built areas, like Sofia and Paris, much of the centuries-old and overlapping infrastructure below their streets—from gas lines to electricity—is poorly mapped, making any system-wide overhaul outrageously costly. Digging up every street to collect wastewater and stormwater in separate pipes could bankrupt these cities if they ever managed to complete the projects.

To analyze the best way to mitigate the worst effects of their CSSs, these cities first need to take a page out of David's book and collect data. During massive storms, measurements of how their systems operate—and where they fail—will determine the most cost-effective places to invest. Authorities in Bulgaria's bustling capital city of Sofia launched such an effort, seeking to make sense of their labyrinth deep beneath the pavement.

There, a team of pioneering DWC researchers collaborates with the city to create groundbreaking, low-cost temperature sensors to help tackle overflowing sewers' devastating impacts. Their first challenge is to minimize how much raw sewage they release into the nearby Iskar River without triggering backups into homes. To tackle this, the sewage project manager for the city's water supplier, Valentina Dimova, is learning how to optimize the raising and lowering of Sofia's 235 overflow gates.

"CSOs are like gates between the sewer system and the river," she explains, "for decades, we have failed in our role as gatekeepers." Now, armed with the data from sensors in CSO pipes and the rivers they empty into, gatekeepers like Valentina can reduce to a science the hard choices that were too often made based on intuition. They can see for the first time how their levers affect the whole system and measure the contamination effects they have on rivers to ensure compliance with the EU's increasingly strict regulations. Like a sentry keeping watch over the castle walls, modern technology has allowed for unseen monitors who can keep vigil and guard against threats.

In Berlin, by combining the same low-cost sensors with AI-driven models, the city can simulate their sewers' activity in software with remarkable accuracy and precision, forecasting future events and testing different approaches to managing them without putting their environment or residents' homes at risk. The modeling also helps them target the best solutions by demonstrating the potential operations benefits of relatively inexpensive tweaks, such as altering the sizes of particular pipes. The practical approaches that Valentina and her colleagues are testing in historical cities like Sofia and Berlin show cities across the world that we cannot be paralyzed in the face of monumental infrastructure problems that cannot be solved all at once; instead, aided by new technology, we can make major strides toward protecting the people, property, and environment around us.

As in Chicago, adding natural elements will be another important saving grace. The EU suggests that we could halt a fifth of combined overflow by covering 17 percent of the nonporous materials that dominate cities with plants—namely concrete, asphalt, and buildings. Dr. Emanuele Quaranta, a hydraulic engineer, and his colleague, Dr. Alberto Pistocchi, an environmental engineer and land planner, reached these conclusions in a 2022 report for the European Commission's Joint Research Centre. They found that the sensor-driven work that Berlin and Sofia now pursue offers significant benefits, reducing CSO overflows by three times the amount of traditional approaches that rely only on growing the storage capacity of water treatment systems. But the biggest benefits, they discovered, come from expanding "green" prevention strategies, which offer tenfold returns.

Critics of urban greening sometimes point to an apparent tension between projects championed in the 2022 report and cities' intensifying housing demands. While certain interventions such as bioswales and rain gardens require space that might otherwise be zoned for apartments, proper greening aligns its flood-resilience goals with a region's housing needs. A house is of little value, after all, if one needs to flee it periodically to escape floods—especially when flooding becomes so prevalent that the property is uninsurable. By managing flood risks that would otherwise leave an increasing number of neighborhoods uninhabitable, greening projects can secure urban housing and facilitate development. When executed effectively, relatively minor investments save cities astronomical costs imposed by major floods.

In light of this, why have so few American cities followed Chicago's lead? As with many of the costs of climate change, it is possible that city officials do not want to recognize the risks until they are undeniable. Others may view green infrastructure as infeasible, not realizing the innovative ways that cities can build flooding buffers without constraining future development. When there is no longer room to preserve larger natural areas, interventions like vertical gardens and green roofs can mitigate flood risks. Permeable pavers can also significantly reduce stormwater runoff without allocating any new space for nature. One solution could be to implement a mandate that requires new developments, such as apartments, in flood-prone areas to incorporate a certain amount of rain-absorbing greenery. Though we must consider the density of urban environments, combined with the benefits of a smart CSS, these improvements could be mighty, indeed.

In Brooklyn, I reconnect with Adrian Benepe, who has tested the power of this synthesis. The former parks commissioner for New York City, he had worked closely with Fiona Watt, New York City's chief forester, during his tenure at the Department of Parks and Recreation. Together, they had sought to make the case for the city's massive greening effort in the early 2000s. Since then, Adrian has moved on to the Trust for Public Land, where he spearheaded the City Park Development Division. He took his love of parks advocacy to the national level by championing the "10-Minute Walk" movement—a cross-country initiative encouraging cities across America to ensure every citizen has access to quality green space within walking distance. After eight years in that role, Adrian was ready for a new challenge. In September 2020 he found it and joined the Brooklyn Botanic Garden as its new president and CEO.

Founded in 1910, the Brooklyn Botanic Garden is a lush 52-acre paradise of over 14,000 plants adjacent to Brooklyn's Prospect Park. It has become an urban oasis for nearly one million visitors who explore its exotic flora and fauna each year. But Adrian was drawn to the role of overseeing the park because of another kind of potential he saw, one that too few green spaces around the country pursue: beyond offering an inviting respite for diverse wildlife to flourish, parks and gardens around the world can serve as powerful blueprints for managing stormwater runoff through innovative water-absorption techniques. With New York City getting battered by stronger storms every year, resulting in floods that devastated entire neighborhoods as recently as 2023, bringing these techniques to green spaces could not be more important. Adrian tells me he was drawn to the challenge of maintaining century-old botanic collections while transforming them into sponges that could save his neighbors' homes.

He saw the need for water absorption across the city two months after retiring from the Parks Department when Hurricane Sandy slammed New York City's five boroughs. The metropolis sustained up to $19 billion in damages, facing the impossible task of processing five billion gallons of diluted sewage in its centuries-old CSS. The results were predictable—entire blocks of buildings were destroyed, miles of subway tracks and stations needed redeveloping, swathes of basement backups, and sudden spikes in dangerous bacteria in the water. As disturbing as it was, Adrian knew Sandy was only the rehearsal for what nature could do, turbocharged by a changing climate. He had to do something to keep more water out of New York's antiquated sewers during heavy rainfall, and his job at Brooklyn Botanic Garden eventually allowed him an opening.

He meets me on a wood-chipped path in a Japanese-influenced section of the Botanic Garden, wearing a tan raincoat. Looking up at the darkening sky, I realize that my linen blazer was a terrible choice. He shakes my hand with firm assurance, half-hugging me with the other like we are old friends. Seeing how he interacts with others as we make our way around the grounds—not just his colleagues, but even tourists who recognize his uniform underneath the raincoat and stop him to ask directions—I realize that his friendly confidence puts everyone at ease.

Meandering through the garden's paths—a range of porous surfaces from gravel to deep-red sand—we walk toward the pond he wants to show me. "Almost every time it rains, New York has a CSO incident," Adrian says. "Which means we're letting raw sewage out in the harbor and rivers, violating the Clean Water Act." Now, as the president of one of the region's largest urban gardens, which relies on a precise balance of water and nutrients to sustain its diverse plants, sewer overflows terrify Adrian. "It's a precarious situation," he adds, his brow furrowing. He tells me that just one-tenth of an inch of rainfall can cause a CSO. A bad enough event could wipe out almost every organism in the garden, permanently incapacitating his organization.

In the years since Sandy, the Brooklyn Botanic Garden built a solution—a pond run by cloud computing, which doubles as water storage for the surrounding neighborhoods. At the bottom of the pond is a smart valve, which monitors weather conditions in real time. The water depth in the pond is allowed to reach two to three feet before the system alerts the pond valve to release water, draining slowly through the rest of the garden. Suppose the weather forecast calls for significant rain. In that case, the valves underneath the smart water garden automatically release the necessary amount of water in the days ahead of the storm, which will keep the pond's water levels steady through the downpour. Through this optimized process, the smart water garden doesn't just stem Brooklyn's floods, removing 52 acres of discharge from the city's sewage system, it also doubles as a reservoir to nurture Brooklyn Botanic Garden's thousands of thirsty plants during the summer heat.

Adrian leads us to a bench assembled out of a nest of pruned tree branches, where we can look across the water to the trickling stream on the other side. The pond's placid surface is broken only by the rhythmic dance of water striders pulsing across the surface. The deep chirp of a northern cardinal transports me to my childhood home in Ontario, where the same birds once ranged across the forest behind the neighborhood.

"I've learned to trust this system," Adrian says, "but it wasn't easy for me."

In 2021, it turns out, the pond was put to the ultimate test. Heavy rain and expected flooding meant that the conventional wisdom suggested the pond should be drained in anticipation, even though New York's sewage system was strained before the heaviest rains crept into the forecast. But the weather-prediction algorithms decided against opening the valve. Adrian admitted that in the worst moments of the storm, he regretted involving himself in the park's drainage system.

A few days later, when the rain came, the pond came within a few inches of overflowing. Then the storm abated. The pond gradually drained its excess water in the days afterward, with the surplus not absorbed by the garden seeping into a municipal sewage system that no longer overflowed. The algorithm had made the right decision.

Every time CSOs happen in New York, the city pays enormous fines to the EPA to compensate for its breach of the Clean Water Act. To save money and the environment, the city has long invested in green infrastructure to capture stormwater runoff before it hits the sewer. That thinking predated even the Bloomberg administration under which Adrian worked. But during his tenure, Mayor Bloomberg created an additional $1.6 billion fund to intercept further stormwater runoff, some of which was allocated to help build this smart water garden.

We follow a meandering path around the pond. I hear the soothing trickle of water over rocks, admiring the lush flowering plants along its quay. Except for the fading roar of a private helicopter overhead, it's a haven I've experienced in a city like no other. Many species planted here, like black tupelo, are resilient to both sodden and arid climates so that the tranquil pond's banks can ebb and flow without damaging the habitat. Among them are moisture-loving sedges and rushes, shooting from the ground in vibrant hues of green. Standing in this serene space amid the calming babbling of water and rustling trees, it's mind-boggling to think that a bustling main road lies just steps away—its traffic barely discernible.

"The estimate is that we're keeping five million gallons of water out of the sewer each year," Adrian says, reminding me why I'm here. The garden's water needs are met through this system, which operates as a closed loop. In New York, where every drop of water supply—from fire hydrants to sprinkler systems—is drinking-water quality, taking pressure off the treatment plants is crucial.

In terms of city-wide stormwater runoff, the smart water garden is a drop in the bucket, just a few million gallons compared to the five billion that the city reckoned with during Sandy. But on a hyper-local level, its effects are significant. Since the pond's construction, the reduced flood risk of nearby streets and homes has put residents at ease. If each neighborhood could have a pond like the Brooklyn Botanic Garden's to nurture water-absorbing nature along its banks and hold and release water at exactly the right moments, cities would never worry about flooding again.

"Shoot—the time!" Adrian gasps. He is late to an event where he's speaking. Adrian's life as president of Brooklyn Botanic Garden has become a flurry of rallying for funds and courting politicians. We jog to his office, and while he changes clothes, I wait outside the building, digging my face into what the sign tells me is a—*Ferraria crispa*, or starfish iris, an eye-catching flower native to South Africa. Its blooms are the star of the show, with six creamy white petals with ruffled edges and mauve spots across its surface. The strong scent of vanilla surprises me—as does Adrian, who, now dapper in his suit, asks if I'm ready to go. We hop in his golf cart, zoom out of the back entrance of the garden, and speed walk to the subway. Eager to maximize my time with the busy man, I keep asking questions, shoving my mini-mic into his breast pocket.

I dread to ask what I've been wanting to all afternoon. But I know I may not have another opportunity. So, standing on the subway platform, I phrase the question: "Do you think New York is ready for the next Sandy?"

Adrian raises his eyebrows and swallows. "As a whole, we're doing a better job," he says. "But no, we're nowhere near ready." The conversation continues onto the screeching subway, which has pulled into the station. "Big ideas were there before, during, and after Sandy. Though there's more momentum to build them now, many are still in progress."

"Like . . . ?" I push. Eleven stops left.

"I think the city made a massive mistake at East River Park," Adrian offers, paying no mind to the curious commuters around us. He describes the rushed, three-week obliteration of a much-loved 57.5-acre, eighty-two-year-old waterfront park on the Lower East Side of Manhattan. Amid a flurry of protests and widespread condemnation, the hasty demolition sparked an outcry from those who felt their—and nature's—rights had been violated. Guarded by police protection, bucket trucks and chainsaws ravaged the landscape, downing nearly a thousand centenarian trees, their remains now buried eight feet deep in a blanket of landfill. Hawks and pigeons desperately searched for somewhere to settle amid the transformed terrain below. Bewildered chipmunks flung themselves from their winter nests in shock. Most perplexingly, the city justified the action to guard its citizens against storm surges by dismantling the park to build a new one on top, eight feet higher above the East River. As the city explained, the new park would transform into a flood barrier, adopting the traditional approach to flood prevention that sought to make neighborhoods immune. The original park, however, amid a gradual renovation that had wide support from the community after four years of public input, also promised protection against flooding. It used a modern approach that prized resilience over efforts to build immunity that the changing climate too often rendered obsolete before they were even finished. By elevating some areas, constructing berms and marshland, and installing protective floodgates on the road that separated the park from the rest of Manhattan, the original renovation promised to create a barrier to the nearby community and critical flood protection between the park and the city.

By dismantling this massive storm-surge barrier for a decades-long project, Adrian explained that the city sacrificed a necessary buffer to prevent the next Sandy from being as damaging. Ultimately, the city said that the original plan was scrapped because installing the floodgates would disrupt traffic.

"My stop!" Adrian points to the door. "Let's keep in touch," he yells over the shriek of the halting car. I slip my recorder off Adrian before he disappears behind the closing doors.

I get off at the next stop to explore Brooklyn Bridge Park, another "sacrificial" piece of green infrastructure designed to flood to absorb water and spare the inland neighborhoods from the worst. Nestled below the hundred-year flood line, the park is fortified to contend with even the most powerful inundations, with flood-resistant plant species planted along each path. Like the hiking trails along the Meuse in the Netherlands, it is a park that looks like nature but works like a dike.

As I wander through the park, taking in Manhattan's skyline at dusk, I stop to wonder what the green infrastructure of the Brooklyn Botanic Garden and along the city's coastline means for my fellow 4.4 billion urban dwellers, with so many of us living in areas now deemed "flood prone." I think about the record-breaking rainstorms that hit Henan province in central China in July 2021. Alarming footage on social media showed the water level inside a Zhengzhou metro station higher than passengers' heads, killing at least twelve as they commuted to work. At the same time, my friends in the east of the Netherlands were desperately working through the night, piling sandbags to protect their homes from the swelling Meuse, the river barely able to hold the torrential rain even after the tens of billions that the country had spent in successful flood-prevention projects. Older folks along the Meuse remember calmer times, where flooding was a distant threat, occurring only once a generation. Now, younger residents like my friends are traumatized by recent events, bracing themselves every spring and summer when they hear the crack of a thunderbolt with a "go bag" packed in the closet, ready to evacuate at a moment's notice. All of us have good reason to live with these fears. The 2022 report of the Intergovernmental Panel on Climate Change (IPCC) has revealed that heavy rainstorms will only become more common and intense due to higher temperatures. Flash floods will no longer happen once a generation but likely yearly.

We cannot stop the rain; we will not stop the storms. No matter what we do, the waters will sometimes rise to new heights that we could have never imagined a few years ago. Instead of trying to do the impossible and bend nature to our will, the solutions that are now most vital must flex with the rain, leaning on the IoN to guard against the worst-case scenarios—mass evacuations, trillions of dollars in damage that leaves cities unable ever to recover, and even deaths. We are no longer planning and building infrastructure for the future but safeguarding our lives and livelihoods for the coming months. Rather than focusing all our efforts on "flood-proof" plans that are destined to fail, wasting years of progress, we must build communities where residents have peace of mind, knowing the torrential rains will infiltrate and nourish nature—the lifeline keeping us physically and mentally afloat.

I called Singapore home in 2013 while I was a college exchange student learning about tropical biology and conservation. Singapore, among other quirks, is known for its regulations that surprise tourists and residents alike. There are rules against leaving the toilet unflushed (officials conduct random checks in public restrooms), and "public indecency" such as walking around naked in one's own home or even connecting to someone else's Wi-Fi—all of which have corresponding fines and possible prison sentences. The city-state is deeply committed to preserving its pristine environment, strictly outlawing the feeding of pigeons, littering, and even banning smoking in public areas. Selling chewing gum—or worse, spitting it out on the street—will set you back S$100,000.

Besides its oddities, the feature that most struck me was Singapore's juxtaposition of modernity and nature, where sprawling skyscrapers meet lush rainforests, and the sound of rumbling engines gives way to the soothing chirps of cicadas. This city-state of nearly 5.5 million offers one-of-a-kind attractions that seem to defy the laws of nature, with greenery pouring out from even the busiest areas. Concrete buildings snugly coexist with tropical trees, winding rivers blending into colorful streets, and sparkling waterways reflecting the vibrant cityscape. Years after I left, this fragile balance would be tested in ways I hadn't imagined.

An emerald metropolis of grass and steel, Singapore has a wealth of green spaces in city parks, wildlife preserves, and its spectacular Gardens by the Bay, a nature park spanning 250 acres in the city's center adjacent to the Marina Reservoir. But beneath this almost futuristic yet idyllic landscape lies an environment fraught with danger—every monsoon season brings flooding rain and warnings of rising waters. Surging tides overrun busy expressways, and epic floods course through neighborhoods like roving bandits, cutting off power supplies and—although rare—at times, erasing lives in their wake.

Singapore is so prone to flooding that one of the most-read articles in the city-state—with more than 250K unique views—lists the top one hundred places to avoid during heavy rain. Due to heavy rainfall, high tides, and drainage problems, dangerous floods are common, especially in low-lying areas where many Singaporeans live. While living there, some residents I got to know felt that they had grown accustomed to the floods, taking protective measures such as storing shoes and electrical appliances on racks to prevent them from getting wet. With the floods in Singapore being flash floods that subside within a few hours, people explained to me how they could live with the water.

The summer of 2021 changed that.

During three hours on August 24, more rain pelted western Singapore than the country had recorded for an entire month. Singapore deployed its military to rescue passengers from vehicles and residents from homes stuck in flood waters. Watching the footage of armed guards pulling stranded people from roofs and into helicopters, I wondered about the friends I'd made there, booting up an old phone to track down their numbers.

While Singapore's national water agency went door-to-door to check in on residents, the source of the flash flood became increasingly clear: heavy rain that inundated the drains' capacities. One of the victims of the August 2021 floods, Changi Airport, had pulled out all the stops to address flooding after an earlier rainstorm took out part of its airfield. The airport had invested in a network of nature-based drains, which can be manually opened to release water like a dam. But in August, the overwhelmed airport staff opened the drains too late, resulting in devastating floods. Now, a network of ten solar-powered sensors uses radar to monitor water levels in gutters. Lovingly dubbed "Google Drains," each sensor is also fixed with a closed-circuit television camera to monitor drainage. Rather than running around in a rain-induced panic, operations workers can now watch the growing network of sensors on a digital map, allowing staff to easily identify and deploy drains from the touch of a button—or set them to open and close automatically. The system's success during a 2022 rainstorm in which 5.6 inches of rain fell in a single day shows just how powerfully a nature-first approach can be married with technological assistance to save lives, protect infrastructure, and keep cities running as they should.

Three cities' experiences—one on a massive freshwater lake, one coastal, and one on a tropical island—show how different revolutionary technologies can be applied to address the needs of urban habitats worldwide. They offer hope that as climate change accelerates flood risks, the IoN can keep the water at bay and keep our homes safe from destruction—if only we dare to plan for our inevitable future.

One important question that urban ecologists must tackle when cultivating the right greening intervention for each city is which plants to include where—and how to combine them. The best bioswales, river gardens, and green roofs, for instance, use plants that absorb significant water and thrive in each other's presence and the city's climate. As

Adrian discovered, with the sedges and rushes that line the banks of Brooklyn Botanic Garden's pond, they must also be resistant to dry periods since dead plants won't absorb water.

Finally, given the limited space for urban vegetation, cities should leverage nature toward preventing floods to provide another important benefit: supporting the region's biodiversity, which sustains key features of our habitat. Drawing together a range of species that thrive in the same soil conditions and complement the insects and birds endemic to the city helps to accomplish this important goal.

CHAPTER 6

The Scientists Within Us

We should preserve every scrap of biodiversity as
priceless while we learn to use it and come to
understand what it means to humanity.
—E. O. Wilson

"THERE ARE PROBABLY A hundred species in here."

I find that hard to believe as we peer into a tree pit. "I see only one," I offer.

My guide for today, Professor Menno Schilthuizen, encourages me to look closer. He takes a few steps from the pit, sitting on a mossy street paver and stretching his legs as if willing to wait all day for me to get the right answer.

I'm crouched on my knees, digging my fingers into the small, grassy concrete-encapsulated tree pit. A man running toward Central Station, presumably to catch his train, runs into me, nearly knocking me over, excusing himself as he pushes on. Stopping to count the species in this patch of green is evidently not a socially acceptable activity in the busy square of this small city.

I brush the dirt off the crumpled poplar leaves, then move them to the side to uncover the dry soil beneath. I guess if you consider the

fungi that are probably in here somewhere, you will get up to five or
so species.

"Fungi, yes," Menno says. He removes his glasses and begins cleaning
the dirt from them with his shirt, leaning back and looking at the sky as
if to tell me he's content to be here awhile. "But what about the bacteria,
mites, nematodes, worms, and other insects that call this tiny tree pit
home?"

My eyes turn to the poplar's bark; it's blanketed in moss, another species, maybe three. I hear a crow cawing above me—another one for the
tally.

Hundreds of people, apparently having disembarked from a recently
arriving train, filter past us, swerving reluctantly out of our way.

A great tit joins the crow in the tree's leafless canopy. A silver gull
lands on the sidewalk a few feet away from me.

"Okay." I turn to Menno. "You win. There's like a hundred species
here—just in this square meter."

Menno rises and brushes the dirt off his legs. A hasty pedestrian
whose face had been buried in her phone nearly collides with him.

"Forgive me," Menno says. Then he turns to me and nods, almost
smugly, proud he's shown yet another urban dweller the extraordinary
richness of life that surrounds us—if only we can open our eyes.

The term to describe this variety, biodiversity, was introduced in the
late 1980s by Walter G. Rosen of the National Academy of Sciences,
Menno tells me, when organizing a national forum on the range of organic life worldwide. Aiming to raise awareness about the important
role that all sorts of poorly understood wildlife plays in maintaining
our habitat—everything from keeping in check dangerous algae in the
freshwater where we recreate to creating the nutrient-rich soil that we
rely on to feed ourselves—Rosen addressed his work to people outside of scientific circles. Rosen recognized that the rampant destruction
of wildlife was threatening to destabilize the habitats we rely on. The
intervening decades have only exposed how dramatic the trends Rosen
first noticed are.

As humans rush to live in and around major cities, this new wave of urbanization has resulted in an unprecedented biodiversity loss. Some of this ecological damage is inevitable in densely settled areas, with buildings displacing animal dens and square miles of asphalt and utility ditches cutting off plants' networks. But recent research shows that much of the damage to biodiversity can be attributed simply to ignorance about the natural world. As Menno argues, our inability to recognize nature facilitates our destruction of it. Tragically, we often do not recognize the vital benefits that species offer us until after they have been lost.

For cities, those benefits are manifold, with our declining biodiversity building on itself and ultimately curtailing our habitat's ability to provide all the services that nature does. In the first step of the process, our fraying biodiversity accelerates vector-borne diseases, resulting in mass casualties of plants and animals. Then, with even less biodiversity, diseases move quicker still across the organisms in our midst. Eventually, we lack the nature needed for everything from combating pollutants to fighting heat waves and mitigating flooding.

Seeing the dramatic costs of nature-bare cities on our horizon, researchers like Menno have launched efforts to head off that future, developing tools to demonstrate biodiversity's power to help address many of our urban problems and build momentum for restoring this vital resource. When communities better understand the flora, fauna, and microorganisms in their midst, Menno believes they will actively protect them.

Despite debate among scientists about the scale of tropical deforestation and its effects on biological diversity, the undeniable truth is that human population centers and their surrounding landscapes have suffered immense losses in populations, species, and entire ecological communities. In urban regions of the United States, the only animal species that have survived are generalist ones such as the raccoon, rat, squirrel, pigeon, chipmunk, and white-footed mouse that can adapt to varied habitats and resource conditions. But the loss of animal and plant species is not confined to urban areas in developed countries. In fact, developing cities have suffered even greater hits to their biodiversity. In coastal Brazil's urban areas, for example, less than two percent of the Atlantic forests remain, with thousands of species going extinct. Most of these species were never even documented by taxonomists, and they came from a region with a high degree of endemic species.

The intricate balance of nature can be thrown off by a multitude of overt and covert influences that unravel the intricate web of interactions between species and cause entire populations to vanish from urban landscapes. Perhaps the most overt threat to biodiversity is the destruction of natural habitats in and around urban areas. Urban activities, such as paving over natural habitats and converting them for other uses, lead to the fragmentation and destruction of functional ecosystems. Construction of commercial and residential facilities and extensive infrastructure development like roads and highways replace lush green spaces with man-made structures inhospitable to the plants and animals that once lived there. In addition, misguided pest control programs and activities such as logging and weed eradication have further adversely impacted biodiversity in urban regions. And other activities, including the channelization and impoundment of streams into underground wastewater pipes, the drainage of wetlands, and even seemingly harmless activities like planting exotic trees and shrubs in parks or converting open spaces into golf courses, disrupt the natural balance of biological diversity. These factors collectively drive populations to tipping points they cannot recover from.

Covert negative impacts on urban biodiversity can be traced back to indirect sources. Air and water pollutants from industrial activity and household products like lawn fertilizers have been found to disrupt natural ecosystems. Pesticides and toxic by-products of industrial production, such as polychlorinated biphenyls (PCBs), sulfur dioxide, and oxidants, are prime culprits. In addition, reduced water tables, overdrafting of local aquifers, ground subsidence, and changes in natural groundwater percolation patterns due to wetland destruction and runoff diversion also affect biodiversity in more complex but equally destructive ways. They, too, push native species past the tipping points where they can sustain themselves.

As a result, invasive species—usually far less diverse—fill the vacuum, preying on the few indigenous ones that survive and outcompeting them for scarce resources. Ultimately, the lack of biodiversity in the urban environment facilitates the spread of novel diseases and parasites that can devastate whatever nature we have left—including the invasive species themselves. Left unchecked, the overt and covert sources of our dwindling biodiversity leave our cities with feeble ecosystems that fail to provide residents with the essential nature services we need.

On the other hand, cities with a wide range of plant and animal species are proven to regulate their local climates better, mitigating the harmful effects of urbanization, such as flood risks, poor air quality, and heat-island effects. Furthermore, because healthy vegetation stabilizes soil and increases air humidity, a diverse ecosystem can significantly reduce the impact of catastrophic events such as landslides, droughts, and fires.

An often-overlooked benefit to biodiversity is pollination, primarily facilitated by a wide range of animal species, including bees and hummingbirds. Pollination plays a vital role in ensuring the survival and productivity of various flowering plants. The resulting fruits, seeds, and other food sources are crucial means of nourishment for wildlife and people. This critical process is necessary for the sustainability of agricultural and forestry operations, green spaces, and urban gardens, which are powerful tools for feeding the world's rising population of urban dwellers.

Strong biodiversity can protect us from deadly diseases since food-chain relationships between native species keep their populations healthy and balanced. Predatory fish, for instance, consume mosquito larvae, which minimize the chance of mosquito-borne diseases like malaria and Zika outbreaks. Similarly, skunks' natural inclination to prey on scorpions reduces the likelihood of human scorpion stings.

But as ecologists attempt to build momentum in cities to save our biodiversity before it is too late, they are increasingly encountering a new obstacle: they cannot motivate residents to care about what we cannot see. Many researchers are now pointing to a global epidemic known as "plant blindness"—the inability to recognize or notice the plants in one's environment. My encounter with Menno outside the train station was an intervention of sorts—the first time I came to grips with the reality that I, too, was afflicted.

Menno leads us down the busy main road in Leiden, the Dutch city where we'd met. Home to the country's oldest university and largest natural history museum, it is a charming town. Menno, an evolutionary biologist, works at the university and museum. However, his passion for biodiversity has seen him traverse the globe for answers about how and why species evolve, from Greece to Borneo.

Menno's attraction to wildlife can be traced back to his childhood, when he'd grown up in small towns surrounded by countryside, seeking companionship in nature. Like many kids, he had a "bug phase," but unlike most, he never outgrew it. Young Menno's insects of choice were beetles and land snails, but eventually his interests expanded beyond the study of insects to include the variety and interconnection of all plants, animals, and microorganisms on Earth—what we now recognize as biodiversity. Now, he's particularly interested in understanding how species react to changing conditions, such as the evolutionary processes sped up by human influence. This fascinating phenomenon, known as "urban evolution," explains how the stunning diversity of life across cities and suburbs is actively coping with humankind's presence.

Ingenious adaptations are unfolding right before our eyes: Several species of birds have pitched their songs higher than their rural counterparts due to vehicular noise pollution. White-footed mice scurrying through Central Park have grown microbiomes to digest fatty foods scattered around their dens by human negligence. Even Caribbean lizards traverse cityscapes so successfully—a herculean feat—thanks to genetic changes that occurred over time. For Menno to fully comprehend this widespread phenomenon, taking small samples is no longer enough. He has recently begun to lean on the power of enthusiastic citizen scientists who enrich their relationship with nature—often improving their mental health—while furthering our understanding of the life that sustains us in our habitats.

New risks to biodiversity in cities motivate the surging interest in the field of urban evolution. Diseases never previously observed have infected urban species such as raccoons and trees, while other animals, such as coyotes and bears, have adopted increasingly dangerous behaviors. We have learned not to follow the path of historical efforts that sought to tame nature's dangers by eradicating species that destabilized ecosystems and unleashed catastrophic unforeseen consequences. Instead, as we discover alarming new trends in our wildlife populations, scientists seek to identify solutions by understanding the organisms' relationship to their changing environments. Urban evolution reveals how these organisms adapt to artificial conditions, illuminating how they can be protected from the negative impacts of human activity.

A few years ago, Menno and his team created a simple smartphone app, SnailSnap, for studying urban evolution in the context of one easy-to-track-down species, which could shed light on how other wildlife change in response to the stimuli of cities. SnailSnap walks users step by step through how to take and upload images of *Cepaea nemoralis*. This intensively studied land snail has been a model species in evolutionary ecology and genetics for over a hundred years. Using artificial intelligence, Menno's team examined nearly eight thousand geo-tagged photos to answer a simple question: Are garden snails in the city lighter or darker in color than their country counterparts?

But in a time when birds of prey are changing because of ingesting rat poison and shocking stories proliferate about new "super-coyotes" attacking full-grown adults, one question bugged me: Why snails? They seemed among the most harmless creatures you'd find on the lawn, and based on their prevalence on my stoop after every summer rain, they didn't appear to be threatened. As it turns out, garden snails were the perfect test case for several reasons. They're found almost everywhere in the Netherlands and live in many locations, from gardens to forests to parks to weeds between cobblestones. They are not too small and therefore easy to find, yet they also can't run away if you want to photograph them. The primary determinant, however, was something else. Their genetics can be readily identifiable: the color of a garden snail's shell results from its DNA, meaning that you only have to look at the color of its house to know what lies inside.

Menno already knew that garden snails adapt to the climate in which they live. He had seen that in warmer countries, the environment resulted in a lighter shell, which would absorb less sunlight. But Menno's research revealed the effect could even be seen in colder countries to measure the impact of cities' heat-island effect. As we now know, the temperature differences of the effect increase with the size of the city, making them far from trivial in a large metropolis such as New York City, which often averages 2 to 5°F warmer than undeveloped land nearby during the day, and up to 22°F warmer at night. This can influence our well-being and health—especially during heat waves and especially as the climate warms—but it also affects the organisms around us.

Studying plants and animals adapting to the heat island offers two important insights. It informs how we can lessen stresses on our urban wildlife, allowing for a more diverse, stable ecosystem and all its benefits—cleaner fresh water, fresher air, and even better birdwatching, to name just a few. Importantly, though, it may also someday unlock secrets for managing our own experiences in life. In the same way that scientists experiment on lab rats to measure how variables might affect humans, we can study the fish in our reservoirs and rabbits in our parks to approximate how our urban landscape may be changing us.

It turns out that city snails had lighter shells, making them better able to withstand the changing climate. The lighter shell reflects sunlight, helping the snails absorb less heat and survive the increasingly scorching urban summers. When it comes to our infrastructure—particularly roofs and streets, which are often black—we have much to learn from the snails.

But the most important lessons we might draw from the snails adapting to cities do not involve building better homes; instead they demonstrate how we can help insects, whose carcasses snails sometimes eat. Insects are the most threatened of all classes of animals and perhaps the most essential for our survival.

The risks to insects stem largely from our cultural blindness to them. While they make up over 80 percent of the world's estimated terrestrial species, they are chronically understudied, which goes a long way toward explaining why only 1.5 million species have been named across all animal and plant groups, even though scientists estimate the world hosts eight million. Like most people, biologists mainly focus on the big species—there are hundreds of scientists currently studying one species of tiger but only one scientist for every 1,500 species of insects.

While we ignore this group that represents the majority of our biodiversity among animals, human activities continue to accelerate rates of insect extinction. Our collective actions are triggering a cascade of consequences, leading to the disappearance of insects worldwide, what scientists have dubbed the "great decline." In 2017, a groundbreaking study in Germany uncovered an alarming trend: the average weight of flying insects caught in traps had dropped 76 percent since 1989. A Danish scientist was able to draw equally disconcerting conclusions from his own experiment: driving along the same stretch of road each summer and inspecting windshield splatter since 1997 revealed insect populations declining by as much as 97 percent.

While we've seen populations of larger, more charismatic species like tigers steadily decrease for centuries, insect declines have been unprecedented in their sweeping range and scale, decimating populations in mere decades or years. In 2019, a meta-study by *Biological Conservation* examined species extinction rates across various mammals and birds and found that insects are declining at nine times the rate of other animals. In four decades, insect populations have diminished by 45 percent worldwide. We are wiping out species faster than we can discover and get to know them—and the immense value they bring to the planet.

The decline portends risks for all sorts of species because insects are the cornerstone of food webs worldwide, providing sustenance for creatures from tiny fish to mighty bats. Even humans take part in entomophagy—up to one-quarter of the worldwide population depends on insect-based diets as an important source of protein and nutrition. Dubbed "six-legged lifesavers," insects are a remarkable and invaluable component of nature's balance. Not only do they provide many forms of sustenance, but their predation behavior can profoundly reduce populations of potentially damaging organisms; one wasp alone can remove up to two pounds of invasive pests from a garden as large as two thousand square feet. Spiders' dietary habits also help keep destructive insect numbers down—estimates suggest that these arachnids consume an astounding 400 to 800 billion tons of insects yearly. Ladybugs, especially in their developmental stages, consume over a thousand aphids, even into adulthood, helping maintain biodiversity among insect populations.

Insects are perhaps best known for their vital role as the world's pollinators, another phenomenon our food supply relies on, as plants generally require pollination to produce fruit. Evolving over millions of years, plants and insects have developed unique symbiotic relationships through pollination, which sometimes need the existence of particular insect species. Figs, for example, cannot be pollinated without their partner in life, the fig wasp. These collaborations support 75 percent of crops.

Through their role as nature's mini-cleaners, insects also maintain the delicate balance of our ecosystems. By consuming dead matter, they break it down into nutrient-rich organic material that is then recycled as new life. From blowflies feeding on waste materials like dung and decaying animal flesh to fungi decomposing plant debris, these tiny janitors keep us from being up to our necks in dead matter.

Insect biodiversity has come to be known as the "hammock" of life on Earth because of its integral role in supporting the environment. Although we have harnessed this resource to our advantage so far, there is a significant danger that continued pressure may ultimately lead us into a situation where natural systems are overburdened with negative consequences. Taking too many strands from this ecosystem without replacing them will render the hammock fragile—until, eventually, our ecosystems collapse under their own weight.

The challenges to insect biodiversity are at their greatest in cities. One major cause is the loss of natural habitats to development. As cities expand and the wetlands and vegetation where insects live disappear, many insects find nowhere to breed. Additionally, pesticides used in cities play a big role, killing off beneficial species or reducing their reproduction ability. A third factor is climate change: rising temperatures, which are felt strongest in urban heat islands, and precipitation changes can affect certain species' survival and reproduction. Finally, pollution from artificial lighting can disrupt the behavior and movement patterns of nocturnal species like moths and fireflies. Together, these forces disrupt the urban ecosystem's balance, leaving our flowers without pollinators and our beloved birds without food sources. Therefore, it is crucial to safeguard the insect environments that we desperately need. This entails preserving their habitats and altering our attitudes toward insects and their significance in our urban surroundings.

Menno and I continue our walk through Leiden's labyrinth of canals and quaint cobbled alleyways. There is little nature, it seems, among these narrow streets packed tightly with buildings.

Menno proves me wrong again as he halts and cocks his chin up, his newsie cap and horn-rimmed glasses mere inches away from the mossy facade of someone's house. I step beside him, realizing I only come up to his shoulder. He puts a finger to his mouth, motioning for silence, and points his other index finger to what has transfixed him: a small dung fly perched on the brick. Its vibrant, lemon-hued stripes stand out in sharp contrast against its tawny thorax, like flashes of sunshine amid sullen mud. Its tiny wings sparkle as if adorned with a thousand sequins, and its beady eyes seem filled with curiosity, darting around in anticipation.

Despite its beauty, the fly's diet is less glamorous, consisting mainly of the excrement of other animals, often congregating around manure or rotting organic matter. As scavengers, a single population of dung flies can help clean up vast amounts of organic waste in their environment.

An unassuming yet essential species, now barely visible as its tapered, lanky silhouette takes flight, the dung fly may appear a simple bug that you'd squash if you found it buzzing around your kitchen—yet it has outsized importance. A keystone species, it helps maintain balance in its fragile ecosystem, providing essential nutrients to the soil and keeping out invasive pests that would otherwise dine on the organic matter it takes care of. Although its existence may not seem glamorous, it quietly works behind the scenes to sustain the city's gardens and parklands.

It disappears, likely in search of its next meal, and I let out a long exhale, realizing that I'd been holding my breath. "I'm always happy to see one of those," Menno says, "because it means we're doing well here." As it turns out, the presence of insects like the dung fly provides an indicator, a biomarker even, of the resilience of an ecosystem.

Contrary to popular belief, he tells me, urban environments adjusted based on the human population are actually more biodiverse than they were a century ago when canals and rivers were used as open sewers and dumps for the booming textiles industry whose factories discharged dyes and other chemicals.

"These waters," Menno says, tilting his head toward the canal beside us, "would've been a dead zone." Back then, canals were too polluted for the now ubiquitous grebes and coots to find their food. Air pollution ensured no lichens could grow on trees. No longer are cities a breeding ground for disease, but rather a new standard in progress in public health. As cities became more hospitable for birds, fish, and insects to live, they, in turn, became better habitats for humans, too.

Still, the population explosion across the world's cities challenges this hopeful trend, cutting further into urban species' remaining habitats yearly.

Menno points to a wooden beam adrift in the canal. Once pale and unmarked, its smooth surface had begun to take on a new life; patches of emerald moss had started to grow, glimmering in the sunlight. A curious coot had taken up residence at one end of the beam. Here, at its temporary home, I imagined the coot's life, enjoying silent mornings, dewdrops reflecting off the water's calm surface. Despite not being depicted in David Attenborough's documentaries, this floating beam was just as much a part of nature's tapestry as icebergs where penguins congregate, a reminder that there is magnificence everywhere, seen and unseen. At long last, I understood what Menno had been trying to show me for nearly an hour.

We often overlook nature hiding in plain sight—a weed, an overgrown gutter, a tuft of vegetation next to a drainpipe, all vibrant intersections of life forms teeming with far more than meets the eye. Along the green surface of that trickling drainpipe dance tiny creatures collaborating with fungi and plants to make this city their home. There exists a tendency to view such urban biodiversity as inferior to its idealized rural equivalents, leading to the blindness that I experienced myself. But just like a tiger stalking through a jungle of hanging vines, the vermillion along the bottom of a gutter demonstrates the natural resilience of our world when given the necessary resources for survival. To appreciate—and ultimately preserve—urban ecosystems, we must shift our perspective, teaching ourselves how to recognize and value their quiet beauty. Only when we pause and gaze penetratingly at these unsung heroes can we truly appreciate the interconnections that make this a vibrant home for ourselves and the creatures we rely on.

When we finally recognize that biodiversity in its smallest form is the foundation for preserving both quality of life and resilience within cities, we can understand the importance of finding an objective way to keep track of it. At Naturalis Biodiversity Center, a national museum of natural history and a research center on biodiversity in Leiden, I met someone who showed me how.

Vincent Kalkman is another scientist who never grew out of his "bug phase," becoming an entomologist specializing in dragonflies and damselflies—or "the lithe grace to summer skies." As a boy coming of age adjacent to the Dutch countryside, their iridescent wings and delicate, mesmerizing movements captured his fascination. But instead of charging ahead and adding more to his list of discoveries, Vincent chose a different route, looking to document the life's work of other insect enthusiasts before it was too late. Early in his career, he realized that amateur entomologists who often had extraordinary abilities to identify insects had failed to record their discoveries due to a lack of experience with proper documentation techniques. As a result, stunning breakthroughs within the field threatened to be lost. Vincent devoted himself to ensuring their work was found, heard, and celebrated by future generations.

In 2017, Vincent's mission to capture and uncover the insect world was made even more dynamic with the invention of computer vision technology. He was introduced to the tool by Laurens Hogweg, a recent Ph.D. graduate who used machine learning to identify disorders in medical images. Working in Big Pharma, however, didn't appeal to Laurens, so he'd joined the Naturalis Biodiversity Center hoping to apply the method he'd mastered to a greater cause: the world's biodiversity. To start, he illegally downloaded every plant photo from observation.org, a citizen-science database of over 58 million photos of plants, animals, and other nature observations worldwide, forming the training dataset for his plant computer vision algorithm.

The remarkable algorithm that resulted left the senior botanists at Naturalis stunned. It could do what these experts had spent their entire lives mastering.

While Laurens fine-tuned the algorithm's work on plants, Vincent was reeling. A groundbreaking study had warned of an "ecological Armageddon" after finding three-quarters of flying insects in nature reserves across Germany had vanished over the past twenty-five years, with grave implications for all life on Earth.

Even as Vincent sits across from me in the cafeteria of Naturalis, his lined face is somber, his emotions raw.

"The fact that insect populations were declining so rapidly," Vincent says, "was shocking." He sweeps a curtain of bright blond bangs from his sapphire eyes. "But it was the lack of consistent monitoring tools that shocked me the most."

The German study, and the handful of others that followed it, used a combination of insect traps and weighing techniques to glean insights into the state of insect biomass. These methods came with a steep price tag, discouraging more research into its catastrophic revelations: the research had consumed hundreds of hours of work by experienced professionals, often resulting in notoriously inaccurate measurements that needed to be thrown out despite the painstaking collection procedures. Insect monitoring had always been a challenging research area to fund due to ways in which cultures—even in the scientific community—tended to overlook the species' importance, and it was made an even more daunting task because of the state of the art.

Unlike most of his peers, Vincent wasn't intimidated by artificial intelligence. Rather, he looked forward to its potential opportunities, which he saw plenty of in Laurens's presentation on computer vision-powered plant identification. He immediately enlisted Laurens to create the same kind of tool for insects. But an insect-identifying algorithm trained on crowdsourced insect photos couldn't make an accurate database. Working with crowdsourced imagery has side effects that are difficult to control, especially in the context of animals. Insects, unlike plants, do not stay in one place. Some are more prone to fly away, and the speeds will vary, biasing photos of slower, more impassive insects. Citizen scientists might also be more inclined to go outside and take pictures in nice weather, biasing images on those that also enjoy warm, sunny days.

Given these limitations to crowdsourced data, Vincent decided his team needed an objective means to take the same photos, in the same place, in all kinds of weather, day and night, and across all four seasons. He also needed to build an insect trap that wouldn't kill its target—since doing so would then draw carnivorous insects. Instead, the ideal mechanism would simply photograph insects when they stopped by.

The result was the DIOPSIS insect camera, an innovative solution for monitoring insects automatically. With a digital camera, computer, and battery-powered screen that works like a magnet to attract flying bugs at night, the device uses image recognition technology—trained using hundreds of thousands of photos—to identify the species in its sights. The fully automated system generates long-term data on both the biodiversity and the biomass of insects, which is estimated based on measuring the insects' length.

Vincent tells me the camera took five long years to develop. Only now does he get to embark on what interested him about the project in the first place—deciphering all those terabytes of data. The analysis never would have been possible without AI. By tapping into the infinitely scalable capabilities of machine learning, Vincent and his team can effectively monitor and count various species with unparalleled precision and speed. The ability to detect species in real time has allowed them to identify habitat loss and ecological changes and inform conservation decisions. It also foreshadows the leaps that AI will unlock in other areas of urban ecology, allowing cost-effective data harvesting that will someday measure the need for a wide range of natural interventions.

"But," Vincent adds, interrupting my thought, "larger insect populations are not always progress," he tells me, taking a long sip of his black coffee. "For example, if water quality improves in an area, you'll have fewer mosquitoes, which probably indicates *more* biodiversity." Only a well-trained entomologist can make sense of what this data means.

Given that there aren't enough of them to study data across the thousands of cities experiencing biodiversity losses and harvest that data, it makes sense to put AI tools like DIOPSIS to work on the data collection side.

Beyond mapping broad patterns, these devices can also be used for highly targeted research endeavors, such as investigating food sources for black-tailed godwits. With the breadth of highly mobile animals that the device can be trained on, it can collect metrics on a range of biodiversity to help scientists identify and reverse declines before it is too late.

Vincent has been getting a lot of attention when it comes to cities. Several city leaders have expressed interest in a DIOPSIS camera or a small network of cameras. But Vincent is wary that many officials curious about the device may not yet understand what they want to know.

"Collecting data for the sake of collecting data won't work," he says, as only a scientist could. In general, Vincent tells me, cities want to know if the biodiversity policies they have in place are working and, if not, what they should do to change it. Even with technologies like DIOPSIS that facilitate vast data collection, the research necessary to verify progress will take time. Working with an insect camera, there is not yet archival data to compare results to, meaning that the first experiments may take twenty to thirty years. So if cities want to make the right changes—ensuring they don't unintentionally hurt other elements of their habitats—they are right to want to get started. Still, they also cannot let their urgency get the better of them.

Once a few cities in each region establish baselines, their neighbors can leverage that data, unlocking exponential benefits. Vincent estimates that the region could build a comprehensive dataset with at least one hundred cameras deployed across the Netherlands—or five thousand across Europe.

Whatever scale he achieves for deploying DIOPSIS, its granular measurements, and future AI-driven techniques will enable cities to overcome one of the core obstacles that has plagued biodiversity work: the inability to measure progress objectively. Nations and cities funnel hundreds of millions of dollars into biodiversity projects every year. Still, without a way to prove the resulting ecological benefits, they are often cut before residents begin to feel their effects. DIOPSIS will give local authorities the information they need to prove the incredible value of targeted investments and maximize cities' outcomes.

Vincent takes a swig of coffee, nearly flipping the cup as he drops it. "I feel hopeful," he says, his downcast eyes betraying doubt, "because we have good technology. It will allow us to catch up to where we need to be. The problem, though, is that we're so many years behind in collecting data."

As I sit outside Naturalis, reflecting on my conversation with Vincent, my eyes dart to a small pond across from me. In it, a bird I don't recognize swims almost gracefully, its head tapered back with majestic crown-like golden feathers. Although I work in urban nature, my bird identification skills are embarrassingly poor. So I did what I always do: I pulled out my phone and took a picture with EarthSnap—which can identify over two million plants, animals, bugs, and—luckily for me—birds.

Within seconds, an answer: a great crested grebe. My heart sinks a little as I realize it's one I should have recognized—a ubiquitous bird not just in Leiden but across urban Europe. Still, I take solace in knowing that my sighting of this bird, common though it is, will be logged in the EarthSnap database, one more geotagged observation that helps us better understand the great crested grebe's habitat. After all, Menno and Vincent introduced me to the enormous power of species data. Menno had shown me how crowdsourced imagery could teach us about the evolutionary patterns of uniquely urban species; Vincent demonstrated that identifying trends in the species population helps solidify the best ways to allocate money to protect biodiversity efficiently.

The biggest problem remaining was one of scale. We needed a much richer dataset to unleash the full benefits of urban biodiversity.

Collecting that data is EarthSnap's mission. But it wasn't how the company was started. In 2012, Eric Ralls—the entrepreneur who founded EarthSnap—was in his friend's backyard when a flower caught his eye. An avowed nature lover who was born, ironically, into a Texas oil family, Eric couldn't identify the flower, and neither could his friend or even Google. Carrying around a library of botany books wasn't an option either. So Eric built PlantSnap, the first iteration of his app. Ten years later, it became the world's most popular tool for plant identification, with a database of over 650,000 plants, 475 million plant images, and over 53 million app installations.

Like with many other IoN innovations, a group of people who appreciate the beauty of ecology provided the basis for the technology. That's because the app relies on a database of hundreds of thousands of photos—depicting flowers, cacti, succulents, and mushrooms contributed by users worldwide. They resulted from Earth.com, a website Eric had launched many years before as a platform for exchanging biodiversity photos. When breakthroughs in computer vision made algorithmic identification possible, Eric leveraged the website's vast library of photos to craft a social networking app that can now automatically identify hundreds of thousands of plant species.

A recent partnership with Snapchat brought the app into the purview of younger generations, which now records plants on the platform in more than two hundred countries daily. As the app began to take off, Eric had another idea, building from his original vision of Earth.com, which was to create the world's largest biodiversity database.

The result was EarthSnap—a first-of-its-kind mobile app where users can capture images of plants and animals and acquire immediate information on species type, environment-related details, and global population concentrations. The genius of EarthSnap is how it leverages an identification tool that users like me love for increasing human understanding of the world's biodiversity: while supplying EarthSnap users with sought-after facts regarding animal movement tendencies and changes in habitats, it collects the identifications that users make—

each of which is tied to its geolocation—to create an open dataset that sheds light on plant and animal behavior. Often without even realizing it, nature lovers are now contributing to the world's largest database of "biodiversity big data," ready to answer crucial questions about climate change. The results of users' observations reveal how growing seasons change in various climates, where certain species are going extinct, and where new, perhaps invasive species are taking root.

The impacts of the citizen-science revolution are already clear, with data collected by amateurs now approaching expert observation as the leading source for global research. Nearly half of the data within the Global Biodiversity Information Facility, the world's foremost repository for biodiversity data, now comes from volunteers. Synthesizing the information from passionate users of EarthSnap and other tools like it, municipal leaders can observe how species like garden snails are evolving to adapt, inspiring new approaches for improving our resilience.

The databases of the citizen-science revolution also highlight what we do not yet know. Democratizing biodiversity data collection has laid bare the glaring data gaps in poorly sampled regions—namely, Africa, Asia, and Latin America—where limited research done by experts has left millions of species understudied, and citizen science has yet to fill the gap. But trends suggest that this may be temporary—as smartphones proliferate, undaunted amateur explorers may equalize our knowledge about flora and fauna worldwide. According to social science research, China, Saudi Arabia, Malaysia, Brazil, and South Korea currently house some of the most smartphone-addicted people. As some of the most biodiverse countries on Earth, they are likely headed for an inflection point, where the rising popularity of citizen-science apps sets off an avalanche of insights into the earth's vanishing mysteries.

Through nature-based citizen science, humanity gains more than just valued data; we can see a path to improving well-being and connection with nature. Citizen science is becoming an increasingly popular way to unite people from all walks of life in a communal effort to understand and protect our environment. Studies have shown tremendous benefits of citizen science, including improved well-being, increased nature connectedness, and pro-conservation behaviors. Participating in citizen-science activities such as bird watching, plant observation, water testing, or photographing wildlife is especially effective for reducing stress because it ensures users actively engage with nature.

When I reflected on this with Eric, he felt that helping others learn to love flora and fauna was crucial to the mission. With the app's recent integration with Snapchat, what was at first a boon for curious nature lovers like Eric now promises to reintroduce nature to those who don't appreciate its marvels. Gen Z, which comprises most of Snapchat's 229 million daily active users, suffers disproportionately from plant blindness. While overlooking natural beauty may sound trivial, it lies at the center of some of the world's greatest crises. Public health experts have pointed to plant blindness as a major contributor to obesity, mental illness, and other challenges; environmentalists have connected it to disastrous environmental consequences because it ensures that we do not notice the alarming changes surrounding us.

With over 90 percent accuracy and useful descriptions of animal and plant behavior post-identification, technologies like EarthSnap democratize knowledge about nature's wonders. But they are not infallible, leaving users still scratching their heads with respect to some 10 percent of species logged into the app. Vincent discovered the same thing with DIOPSIS, which, despite advances in AI, stubbornly refused to identify a small portion of insects.

The organisms that apps fail to identify show the key role that experts must still play in a citizen-science-driven world. The principal reason that AI fails to identify a species relates to its training data's limitations, which do not include unidentified species or have been studied so little that experts record insufficient photographs. The need for expert research only increases as more undiscovered species are found.

Back at Naturalis, two cybertaxonomists, Jeremy Miller and Oscar

Vos, are trying to close one of these gaps that no doubt surfaces on identification apps, photographing seventy of the rarest beetle species on
Earth, which—to the untrained eye—look almost identical. Taking taxonomy into the digital age, cybertaxonomy, species identification that
relies on the internet, shares the traditional goals of taxonomy, aiming
to discover, characterize, name, and classify species. But unlike earlier
iterations of the field, its use of cutting-edge technology facilitates far
more efficient ways of analyzing organisms.

Combining cybertaxonomy with the potential that citizen science
apps offer for scaled observational data, scientists can identify new organisms, reveal their phylogenetic relationships, and map out ecological
distributions globally faster.

Beyond photos, some citizen-driven apps use other stimuli—namely
audio—to identify species that may be hard to spot. This is especially
valuable when it comes to birds. Since 2002, the Cornell Lab of Orni-
thology's eBird platform and accompanying Merlin app have harnessed
the collective expertise of amateur birders worldwide, making it available for scientific research, conservation initiatives, and educational
purposes. The platform provides innovative tools that help record observations with photos, audio clips, and real-time species distribution
mapping. There are also alert features, notifying nearby users when
certain birds have been seen in a region. As of 2024, eBird is among
the world's largest biodiversity-related science projects, with more than
100 million bird sightings contributed annually by eBirders worldwide.

Because birds prove difficult to approach without scaring them off,
eBird has trained its algorithm to recognize species based purely on
birdsong. Identifying birds by their singing has always been a challenging feat, even for humans. Still, Merlin has proved that trained on a
large enough dataset, computers are up to the task of pinpointing the
exact melodic range and vocalizations with regional nuances of birds.
Certain species, like mockingbirds, even improvise their tunes like an
accomplished jazz musician. Whereas other birdsong-recognition apps
had tried, offering only a 50 percent accuracy at best, Merlin's latest

iteration achieved much greater results using an eBird database to which tens of thousands of citizen scientists contributed notes and recordings. It boasts near-perfect identification rates for more than four hundred species from the U.S. and Canada. The secret sauce lies in *how* Merlin recognizes birdsong, not through the audio recording but by converting birdsong into a spectrogram—a picture of sound. Those pictures then undergo computer vision training just like photos submitted to SnailSnap, DIOPSIS, or EarthSnap would. It is yet another example of how the power of AI opens possibilities for supporting urban nature that would have seemed impossible just a few years ago.

Building on the hundreds of millions of observations available through eBird's public data, several additional tools illuminate specific features of bird behavior. For example, the BirdCast tool offers a unique view into bird migration patterns, allowing users to track species migration patterns as they occur.

This data is precious for protecting biodiversity, given the danger that tall, illuminated buildings pose for migrating birds. As many cities have realized, the bright lights on their buildings disrupt avian orientation mid-flight, confusing and exhausting birds and making them vulnerable to collisions with windows. Why artificial light disorients and attracts birds is still unknown, but its massive death toll is abundantly clear. Up to one billion birds, including several high conservation concern species, are killed annually from direct collisions with illuminated buildings, towers, and other structures across the U.S.

Scientists fear the problem will only get worse. When birds are in flight, they make "flight calls"—vocalizations that keep flocks together. When disoriented individuals make flight calls, they can lead other migrating birds to artificial light sources, spawning a vicious cycle of increased mortality rates. In 1999, Chicago, situated along a major north-south migratory bird flyway, was the first American city to raise awareness of the problem lights pose for birds in an urban center. Together with the National Audubon Society, a nonprofit environmental organization dedicated to conserving birds and their habitats, the city established the first Lights Out program, which educates the public on dimming lights to protect migratory birds. Since then, groups in over thirty other cities have organized similar programs.

Citizens who sign the pledge are encouraged to help put an end to light pollution's harmful effects on birds by keeping the nighttime sky by turning off spot and decorative lights when not in use, closing blinds and shades, making sure outdoor lighting is down-shielded, and utilizing timers or motion sensors. Unfortunately, the burden these programs impose on normal life has limited adoption. BirdCast, a new app that offers a targeted Lights Out Action Alerts program, promises to make it far easier to save birds' lives. By analyzing bird migration traffic detected by radar, BirdCast offers live and local bird migration alerts throughout the continental U.S. Using this tool, you can determine whether birds are migrating in your area on a given night in low, medium, or high densities, providing timely information about peak migration nights, which occur only a handful of times throughout a given season. This allows effective targeting of Lights Out interventions. Like the best IoN technologies, it helps us protect the nature we love without asking us to compromise the modern comforts that we have also come to appreciate.

A major part of the journey to restore biodiverse habitats involves rehabilitating a relationship in which we invest too little energy. Like jaded couples who take their partners for granted, we have too often allowed ourselves, ensconced in modern comforts, to grow blind to the abundance that nature provides in our lives. Our transformation as a species from nature's hypervigilant participant—at one time able to read the sky of an oncoming storm and hear the distant wail of a predator—to its inattentive boyfriend or even its thoughtless boss is evident in all the ways we have withdrawn from natural spaces. We have lost touch in too many ways with the biodiversity and spectacular beauty of this planet we call home.

Yet the benefits of the technology that at other times isolated us from our surroundings can help us rekindle our connection with nature in exciting ways. We can pool our observations about wildflowers and butterflies to make even a short lunchtime walk an experience that recenters us in our habitat. We can fulfill our social needs simultaneously as we do our elemental ones, connecting with our neighbors to observe a storm of birds migrating across the cloudless sky.

Merely recognizing the multitude of plants and animals around us through an app may not produce meaningful change. Still, it could facilitate a much more profound connection, a deep-seated love for biodiversity. By connecting this emotion with modern technology, anyone with a smartphone has the potential to become an ambassador of conservation on a global scale. In this way, our appreciation for the wildlife that our daily lives too often lack could change the course of our relationship with nature—from a story of devastation to one of mutual support.

It is a vision that echoes one Menno shares with me as we sit on the same bench from which I'd watched the coot drift on a beam down the Leiden canal.

Menno dreams, he tells me, of a daily biodiversity report, like a weather forecast, informing residents on how to make better decisions to protect all the living creatures in their city. "In the future," he says, "each garden, balcony, or even windowsill could be a portal to discovering the wildlife that surrounds us, and we will understand, even from a young age, how to give our surroundings the resources they need to survive." He casts his eyes over the pond that had captivated me earlier. He says we could teach children to be stewards of these miniature habitats, identifying the organisms within them. On a clear day, they can sense what time it is using the sun or envision the crow circling overhead simply by its cackling call. When they venture outside to plant the flowers and herbs they incubate, they can scoop a handful of black soil, see the tiny scavengers dart away from their fingers, and know that this will do.

CHAPTER 7

Immersion Therapy

Come to the woods, for here is rest.
—John Muir

"SHE TOLD ME IT felt . . . invigorating," recalls Blake Ellis, ecotherapy program manager at California University, Chico, or commonly Chico State. "For the first time in three years, seeing a downed tree on the ground didn't bring her to tears. Instead of death and decay, she saw a trunk teeming with new life."

Blake recounts an ecotherapy session she organized for the 2018 Camp Fire survivors. Her patients were edgy and grief-stricken from the destruction they'd witnessed and the instability it imposed on their friends and families. But Blake saw an opportunity to grow from the tragedy that had struck.

Chico State led the effort, bringing Blake and fourteen other local mental health professionals and community leaders to its Big Chico Creek Ecological Reserve, where it trained them to become certified ecotherapy guides. The goal was to comfort the communities most impacted by the wildfire.

I peer over the canyon edge in Bille Park, one of the sites where Blake frequently hosts her sessions. It's the jewel of Paradise, a green expanse nestled between people's rebuilt homes—or in some cases, carcasses of where houses used to be. As I gaze down the cliffside into the vast expanse of the canyon—with layers of green spouting from jagged rock—my breath catches. No wonder people call this the Little Grand Canyon.

I lean forward slightly to see a blue ribbon, the Little Butte Creek, glimmering across the bottom of the valley.

"Careful," a cheery man walking his dog warns.

The man tells us we're among the first to enjoy this new view. The park reopened only a few weeks prior. Once, a wall of trees and brush lined its outer edge along the cliffs, guiding the park's trail toward a lookout point. After the Camp Fire, Bille Park closed its doors for years to remove the burned trees that loomed dangerously over anyone brave enough to sneak past the closed gates.

"It's odd," Blake tells me, "how you can simultaneously grieve the landscape that was there—and at the same time, know it was that same dense brush that got us into this mess." The tug between her heart's nostalgia and her mind's knowledge of the dangers of the brush speaks to the tense equilibrium of the ecosystem, destabilized by earlier generations. Because forest managers and policymakers removed fire from this ecosystem, the forest grew so dense that the fires came back with a vengeance.

As the creek eroded over the course of millennia into Earth's crust, it revealed the ancient cliffs before me, which seem to hold the secrets of time. It's nearly five P.M., the golden hour of the day, and the colors of the crevices of the rock in the cliff across the canyon trace intricate shadows with the setting of the sun, tapestries of red and black that transform from one minute to the next. Once lush forests appear barren and stark in the aftermath of the wildfire. An eerie silence replaces the rustle of leaves and chirps of birds.

Removing trees along the top of the cliff served as a poignant reminder of the devastation that had taken place here. Still, it was also a tribute to our capacity to rebuild in collaboration with nature. The now grassy edge of the park opened space for new growth, an opportunity to build a rejuvenated ecosystem. When I return to this place a decade from now, I know that a flourishing young forest will greet me.

For those who have experienced the trauma of a natural disaster, the thought of returning to nature may seem daunting—even counterintuitive. Yet as Blake reminds me, strolling through the park, it is in the moments of darkness that the healing power of nature can be most profound.

Ecotherapy, also known as nature therapy or forest therapy, provides a unique approach to healing the relationship between humans and the natural world. This method involves immersing oneself in the beauty and serenity of a rich natural environment, allowing nature's sights, sounds, and smells to soothe the soul and ease emotional pain. For those who have lost their homes, possessions, or even loved ones, the healing potential of nature can provide a sense of grounding, one's physical movement a symbol of the ability to persevere through suffering.

And people don't have to go far. Blake tells me about Bille Park, which despite its incredible view, is the kind of urban park that many people can access worldwide, a fine backdrop for many suffering survivors. The park is wheelchair accessible, has plenty of bathrooms, and is more welcoming to those that may feel intimidated going into the backwoods. "You don't have to go deep into the wilderness to connect to nature," Blake says, "my ecotherapy guide loves to tell the story of how he once led a successful session in a parking lot, under the shade of a single tree."

Since she was forced to start her training virtually at the height of the COVID-19 pandemic, Blake did most of her practice sessions in her backyard in Chico's core. "Sure, you contend with construction, honking cars, and pollution," she recalls, "but it's possible to connect with nature everywhere." She frequently ends her walks with a reminder to her patients that nature is always with them—whether a breeze on their face or a glimmer of sunlight through an office window—they just have to slow down and soak it in.

Ecotherapy has its roots in Japan. During the 1980s, the Japanese economy underwent a significant transformation when agriculture, manufacturing, and mining, previously central to economic growth, took a back seat to new industries such as telecommunications and computers, which became increasingly vital. This shift marked a turning point in Japan's economic history as it began positioning itself as a leader in technological innovation and information-based industries. It also created a disjuncture in a society that took pride in its sense of tradition. The growth of Japan's tech industry and the promise of employment opportunities drew large swathes of the population to cities, dislocating them from the land where their families had often lived for generations. While this rapid urbanization propelled the economy, it brought serious social consequences, particularly concerning public health. Among them is a severe increase in preventable diseases such as diabetes, heart disease, and cardiovascular disease. Even anxiety and depression became more common as individuals struggled to cope with the stress of city life.

Faced with new challenges in healthcare and public policy, Japan embarked on a series of research projects that asked a simple question: What happens when humans spend time in natural or forested environments? Their discoveries were profound.

One thing that happens when spending time in a forested or natural environment is that the parasympathetic nervous system is activated, allowing people to recover from stress and adrenaline. As a result, cortisol levels drop dramatically. Other effects result from phytoncides, a type of phytochemical that trees and other plants continuously emit. These antimicrobial, antibacterial, and antifungal pathogens help keep trees and plants healthy and protected, and they also have a positive impact on humans. Exposure to phytoncides has been linked to improved creativity, focus, attention, and academic performance. In addition, being exposed to phytoncides can enhance the function of natural killer cells, a type of white blood cell that plays a crucial role in our immune response against viruses and cancer. A study conducted in 2009 discovered that this heightened activity of natural killer cells lasted for more than a week after the research subjects took a walk in the forest. This indicates that even a brief stroll among trees can have long-lasting benefits for our overall health.

The Japanese called the practice that grew out of its government research *shinrin-yoku*, or "forest bathing." The insights of Japanese researchers now inspire Blake and other social workers worldwide. She and several of the Camp Fire survivors she works with who are also certified as guides build from the science of forest bathing to help themselves and the communities they serve, teaching how one can heal from trauma and reconnect with themselves in a dramatically transformed landscape. As the number of environmental refugees surges around the globe, the work has taken on growing importance.

"It is essential to remember our place," Blake tells me. "As a species on this Earth, who can adapt to different environments. And my place, leading groups toward the benefits of connecting with nature. I am simply a guide. The forest is the therapist."

As the Japanese pioneers demonstrated, being in nature offers an antidote to stresses far less severe than natural disasters—namely, it alleviates the chronic malaise of urban life. Today, urban green spaces are increasingly recognized as a popular way for people worldwide to find respite from modern-day stressors.

The first step toward addressing urban stress involves recognizing its various forms. While physical threats such as heat waves, wildfires, and floods may present the most immediate risks to city residents, the psychological implications of living in concrete jungles are also dangerous. Philosophers such as Henry David Thoreau have recognized the mental health repercussions of living in busy urban environments for centuries, and the COVID-19 pandemic only magnified the challenges. When city-dwellers found themselves confined to their homes without the usual escape of nightlife, travel, gyms, and other forms of social interaction that cities facilitate, it led to profound increases in stress, anxiety, and depression. Then, the closure of some urban parks left certain residents without key outlets to process these emotions.

As we hurtle forward in our fast-paced, tech-driven lives, my conversations with Blake remind me that we must learn to manage the stress of modern life and maintain a sense of harmony with our bodies and minds. They inspire me to wonder whether nature—one of the forces that urban life too often deprives us of—can help us regain our emotional footing by orienting us in our surroundings.

On March 15, 2020, I had just rounded off my Ph.D. fellowship at MIT, and my partner, Robin, and I were supposed to board a flight to Los Angeles that afternoon to kick off an eight-week tour of America's west coast—just us, our rental van, and—for the first time in years—no obligations. My field and lab work were finished, and my data was analyzed. All I had left to do was edit my dissertation. But before I'd do that, we'd give our Boston adventure a proper send-off with an all-American road trip.

Earlier in the week, we'd learned that things were going awry. Four days earlier, in a primetime address, President Trump had announced a thirty-day travel ban on foreign travel to the U.S. from most European countries as COVID-19 cases surged globally. Robin's mother had been visiting a few days prior, and new university rules barring everyone but students and staff from campus had prevented her and Robin from attending my final presentation. Two days later, my parents, heeding new warnings about the risk of contracting COVID-19 in air travel, had driven 900 km from southwestern Ontario to help Robin and me pack up our apartment to travel light to California.

"This will be like SARS," I remember saying to my parents, "local and contained." When they arrived on Friday afternoon, California van life was still a go. "It's the perfect isolation vehicle!" Robin observed. But by Sunday morning, the vacation plan had come undone. State and federal authorities had shut down the local and national parks we'd scheduled to visit, dashing our hopes of reconnecting with nature. Within an hour, we'd canceled our flights, our van, and booked a rental car to follow my parents and all our belongings across the Canadian border. Boston was a COVID-19 hotspot then, and the Canadian border control agent gave us a skeptical glare. But eventually, he waved us ahead. Twelve hours later, the borders would close. Canada would keep its borders closed for 930 days.

For the first week, waking up in my old childhood home, I felt so grateful to be there. Going back to Amsterdam wasn't an option since Robin and I had rented our place for another couple of months. Under the difficult circumstances, we didn't even want to ask the tenants if they'd consider renegotiating the lease.

But after a week, reality set in. This would not be just "two weeks to slow the spread." We were in this quarantine for the long haul. And so, the dread followed. The grief of missing the road trip of a lifetime that we'd felt we'd earned. The anguish of having to get straight back into the dissertation I had hoped to postpone another few months—because what else was I going to do? The innocent woes of being an adult sharing tight quarters with your parents.

In a time when so many media outlets and politicians told me to isolate myself indoors, I felt compelled to venture outside, torn between this elemental desire and the building fear of the deadly disease. As some public health experts noted the exceedingly low risk of catching or spreading the virus in the open air, I felt liberated. I went on long walks in my neighborhood and surrounding forests, reconnecting with neighbors I hadn't seen in a decade, with childhood friends who—like me— had returned to live with their parents, with my family who I wasn't just visiting but living with, and—perhaps most importantly—with myself. Nature became the backdrop for morning meditations, make-shift birthday parties, therapy sessions when friends lost their jobs, and Friday-night happy hours. In many ways, nature was all I had left. What surprised me was how much the outdoors left me with. The words "bad weather" no longer had any power over me. I found myself reconnecting with trees I remembered from my childhood as upstarts only a dozen feet tall, now coated with pillowy moss, with squirrels that no doubt descended from those I remembered springing gracefully along the fence in our backyard. Nature has always played an important role in my life. Still, as I'd spent more time poring over data in recent years, thinking about the natural world only when I put it under a microscope, nature started to feel like the aunt you only see at Thanksgiving. COVID-19 brought her to the forefront again, reminding me how it was the protagonist in the story of my life. It gave me my bearings again and restored a connection I hadn't felt since childhood.

As it turns out, I wasn't alone. In June 2020, Canadian nonprofit Park People conducted bilingual surveys to learn about the experiences of Canadians and municipalities during COVID-19. Out of sixteen hundred respondents, nearly three-quarters of Canadians said their appreciation for parks and green spaces has increased during COVID-19. Almost two-thirds of Canadians reported visiting parks at least several times a week.

Businesspeople building products that facilitated excursions into nature certainly noticed the trends that surfaced in the survey. Ron Schneidermann, the CEO of AllTrails—a curated, crowdsourced trail map catalog—was astonished by the uptick in user activity. AllTrails had been around for a decade, growing to host over 300,000 trail guides in two hundred countries across all seven continents. He came to the company after leading Yelp's business development because he strongly believed in immersing himself in nature. A father of three young children, he found himself frequently using AllTrails' app, aiming to raise his kids "on the trails" as a self-proclaimed "mediocre mountain biker." As people reconnected with the woods around them during quarantine, Ron noticed a massive spike in activity on the app. The number of logged hikes grew more than 170 percent compared to the year before. And not only were more people hiking, but the pandemic also saw an increase in walks per user, with hikers logging 52 percent more hikes in 2020 alone than over the previous four years combined. All of this was great for business, but the trend that made him happiest wasn't the growth—it was where users' growing activity took place. "The most remarkable feature of the trend we saw in 2020—which has continued through 2024," he said, "is *how* people used the technology. AllTrails was no longer a travel app you'd jumpstart when traveling on vacation. It was used throughout the week to discover trails in local neighborhoods."

Dr. Zoë Volenec of Princeton University found similar upticks in urban park usage. She geotagged social media data from state, county, and local parks throughout New Jersey to compare park usage during four discrete stages of spring 2020: before the pandemic began, during the beginning of the pandemic, during the New Jersey governor's statewide park shutdown order, and following the lifting of the shutdown. She found that park visitation increased by 63.4 percent with the onset of the pandemic, decreased by 76.1 percent when the state ordered parks to close, and then returned to the elevated preshutdown levels when closed parks were allowed to reopen. Even as restaurants reopened and city dwellers regained some of the opportunities lost during the lockdown, many did not relinquish their new habit of immersing themselves in the wild several times per week.

The spring sun filtering through the leaves cast lacy shadows on the trillium-speckled forest floor. After last night's rain, the air was thick with the woody scent of wet earth and the pollen of budding wildflowers. Twigs crunched under the rhythm of my mother's footsteps as we walked side by side through the forest. I exhaled, letting my breath slowly whistle through my teeth over the tweeting of birds.

As we approached an opening in the canopy to our left, I tenderly parted the branches of a bush beside the trail to see it. Covering the forest floor were white trilliums scattered like lost pearls. Living through one of the loneliest periods I could remember, I'd struggled to get out of bed that morning, only rising to appease my mother. The flowers reminded me that renewal was not only possible but inevitable. Just as the winter had thawed last month and nature had kicked off its annual cycle of rebirth, we, too, would come out from the pandemic.

The soft scent of pine rushing toward me in a cool breeze suggested that the old-growth forest was not far off. Soon, I thought I heard the gurgle of a stream, and I stopped to confirm that I wasn't imagining it.

"Are you coming?" my mother asked, turning to realize I was frozen on the path.

I nodded.

As I journeyed ahead, a little chipmunk scampered away from the trail, disappearing into the brush. A family of robins flitted from tree to tree, singing softly. I felt like an intruder in their sanctuary, and I moved carefully, doing my best not to disturb the peacefulness that surrounded me.

Amid the quiet was a sense of joy and peace that I'd forgotten, that had eluded me for months during the quarantine. By the time we reached the last few steps, I'd forgotten where I was. My legs were heavy, and my hands warm. Then, the rumble of a car engine reminded me, and a few steps later, we emerged onto the street just two miles from my childhood home.

Eager to add variety to our almost daily pandemic walks, we'd turned to AllTrails to discover a host of new trails. Despite having spent nearly twenty years growing up there, confident I knew my southwestern Ontario surroundings well, I'd stumbled on a path neither of us had ever heard of. The green oasis behind suburban houses reminded me of the unpredictable beauty of natural spaces, which I'd spent years missing. They offered space to refocus my attention in a time of anxiety. After years of thinking of walks as exercise, reducing them to steps counted or calories burned, I now saw their value as meditations and opportunities to de-stress. And I started noticing the moss engulfing the pebbles by the creek, the woodpecker making haste with his dinner, and the flash of a fox's bushy tail between the trees. The pandemic gave me a reverence for the value of a new trail, a small patch of green, or a wide-open space I could visit on a short lunch break.

Whether the average urban hiker realized it or not, we were nourishing a need we all share, which modern conveniences have led us to overlook for far too long. Because of crowding, overstimulation, and a feeling that one is constantly being watched, cities challenge our mental health in ways we are not evolved to handle. The risk of developing depression is 20 percent higher for those who live in cities, the rate of generalized anxiety 21 percent higher, the rate of psychosis 77 percent higher, and so on. The acute and chronic stresses of poor air quality, crime, high costs of living, and other factors propel the brain into a constant fight-or-flight response. Poor sleep brought on by light and sound pollution can further exacerbate mental distress.

At no point in history were these costs more apparent than during the COVID-19 pandemic, when many of the benefits of urban life were canceled—everything from the wide-ranging economic opportunities for those who weren't able to work from home to mass transit and social experiences such as live music, bar hopping, offices with Ping-Pong tables, and disco boxing exercise classes. Forced to stay in or near our little structures, the pandemic laid bare the grim challenges of city life.

As long as humans have crammed themselves atop one another on city blocks, we have understood elementally that nature provides an antidote to urban malaise, even as we have failed to embrace it properly. By the mid-nineteenth century, after the Industrial Revolution brought millions from rural villages to the large cities of Europe and North America, many philosophers and sociologists recognized the crisis and looked to nature to address it. The first parks were built to give inhabitants of these growing cities access to green areas, a much-needed respite from a polluted, agitated, and busy urban life.

The parks were a technology of sorts—carefully cultivated oases designed and cared for to withstand settings that had utterly failed to make room for nature. Yet as we continued to load our cities with other technologies—cars that displaced pedestrians and horses from the streets, steel buildings that reached into the sky, casting shadows over entire blocks, radios and televisions and computers that gave us less reason to leave our homes—we too often forgot to turn our innovative spirit back to the restoring elements of the habitats that our species evolved in for hundreds of thousands of years. The more that modern life changed, the further nature seemed to recede.

Despite the documented health benefits of parks and data on their increased use during the pandemic, many provinces, states, and cities instituted park shutdown orders due to fears that public parks could facilitate COVID-19 transmission. When I returned to Amsterdam in the summer of 2020, I noticed an even more bizarre measure: to stem the flow of visitors, the city had closed certain park entrances and outdoor recreational opportunities like theme parks, zoos, beaches, and swimming pools. The measure was instituted after young people, desperate for social contact, flooded the parks on increasingly warmer weekends. The policy had the opposite effect of reducing crowding: instead of allowing teens to distribute themselves across the city's range of outdoor spaces, it forced all the residents desperate to be outside to line up at the few points of entry and exit where parks were open.

The irony was not lost on Mrs. Temple, my then-ninety-eight-year-old neighbor. As I return to my apartment on a warm July day, lugging the weekend's groceries into the lobby, I find her standing in the doorway, looking out over our street at Amsterdam's Vondelpark. Her lips are pursed, eyes narrowed.

"What's wrong?" I ask.

"Those idiot bureaucrats," she says, shaking her head and fighting back tears. "They closed *my* entrance to the park."

I, too, found myself frustrated at the ridiculous measure, having to walk an additional 250 meters to the nearest entrance, my heavy groceries in tow. I'm about to tell her I can relate when I realize that I can't possibly. I—and the teens and young adults the measure is aimed at—can simply walk to the next entrance. Mrs. Temple cannot. For the entire spring, summer, and beginning of the fall of 2020, Mrs. Temple could only see the park—where she found solace and connection for seventy years—from her living room window.

According to the U.S. Environmental Protection Agency, the average northern European and North American spends 90 percent of her life indoors. Humanity's sudden shift to indoor life, and the resulting lack of sunlight exposure and vitamin D production, is so widespread and so predictive of bad health outcomes that many scientists have called vitamin D deficiency an epidemic. As of 2022, there is ample evidence that higher vitamin D levels are associated with lower rates of COVID-19 infection, hospitalization, and death. Inference modeling and the biological mechanisms indicate that vitamin D's influence on COVID-19 is very likely causal. They suggest that our distance from nature in the time leading up to the pandemic may have helped to spark the deadliest pandemic in modern history.

As the waves of the pandemic crashed, retreated, pummeled, and gave way, I became fascinated to understand if perhaps our politicians had gotten it all wrong. A growing number of mental health experts, grappling with the widespread challenges that isolation had unleashed, argued that lockdowns had disproportionate if unintended consequences. Over four years after the pandemic began, we now know lockdowns contributed to worsening mental health disorders associated with depression, anxiety, stress, diminishing psychological well-being, and even suicide. Yet at the same time that so many struggled through periods of physical inactivity and feelings of loneliness and boredom, the data from AllTrails and other sources measuring nature interactions indicated that quite a few others took advantage of the difficult period as an opportunity to be healthier than ever.

With the benefit of hindsight, I began to wonder if those same devices on which we squandered hour after hour doom scrolling social media, waiting out the pandemic, might offer an antidote to the isolation from nature that the pandemic revealed. Could other technologies, as AllTrails had, help us restore the natural experiences that our urban souls craved?

"Come over here," a soft voice calls out. "Yes, dear. Here, a little closer, can you hear me?"

Standing alone on a street in the neighborhood of Amsterdam North, I swiveled my head, disoriented. I had put on my headphones as I left my house an hour ago, so I should have been prepared for these sounds. But having raced to get here, my mind strayed to the tram chauffeur, who'd admonished me for getting on at the wrong entrance, to the woman who shook her head at me mournfully after I apologized for bumping into her. Now, it seemed, I was hearing voices.

Then, as I looked up, I was grounded by the sprawling oak before me, its pointed leaves catching the breeze and reflecting ever-changing shadow puppets on the ground.

Finally, I remember why I came to this small square, surrounded by misfit houses—some renovated to the tee, others neglected for decades—and a glassy branding and design agency that sounded the pitchy trumpet of gentrification.

"Do you have a second to listen?" the voice asked me.

This time I nodded, fully aware that I was responding to a tree. Based on her warm, kind voice, I assume the oak's a female, and judging by her large crown and wide trunk, she's at least half a century old.

"I'm Marije," she begins, "and I'm so terribly lonely. Two years ago, I could catch up with my friends on the other side of the street. But now . . ." she trails off.

Behind Marije, I see a row of elms neatly planted along the quayside, separated by a busy road.

"When they expanded the road," Marije continues, "the soil got so compacted underneath that my roots didn't stand a chance of getting through to my friends. Now I'm here, all alone."

The tree is one of six "talking trees" featured in the "Giants of the North" immersive, geolocated audio tour by theater director Puck van Dijk. In 2019, Puck moved to Amsterdam North, where she was enamored by the neighborhood's people, culture, and original inhabitants: the oaks, gingkos, elms, maples, chestnuts, and ashes. The product of a nomadic childhood, darting from continent to continent, Puck regretted never staying in one place long enough to know where she truly belonged. When she had a child of her own, Puck wanted stability. She looked to the oldest residents in her neighborhood to establish that sense of place and permanence—her new tree neighbors and the literal roots they'd established over centuries.

The 200-foot-tall plants have seen it all: the collision between a Vespa and a Canta, the marital quarrel in the pouring rain, the lost tourist, the cement trucks roaring by bound for a nearby construction project, their wheels thumping over potholes. Through each monologue, Puck forces you to confront the question: If we listen carefully, what can we learn from these ancient city dwellers we walk past daily without glancing up at their verdant majesty? By taking an interest in these trees, we can foster a deeper appreciation and understanding of the natural world. From marveling at the intricacies of a delicate flower to watching a bustling bird feeder, these small acts can ignite a passion for nature that fuels the desire to protect and preserve it for future generations. Urban nature, which invites us to learn about it, reminds us that regardless of how far we stray from the natural world, we can never fully sever our connection.

Early Monday morning, after participating in the "Giants of the North" premiere, I sat down with Puck to discuss her project. She tells me she'd wanted to be an actress since she was little—until she earned a few gigs and realized that she loved directing theater even more. The immersive experiences a director could create—through technology-aided sounds and visuals—opened a whole new world for her. She eventually became fascinated by how the same techniques could enhance everyday surroundings to create a unique, augmented reality.

"What if these trees could talk?" Puck would ask herself. What would they tell us about their upbringing in Amsterdam's changing northern neighborhood? What would they say about its gentrification, rampant development, and rapid loss of greenery? She had at first tried to understand her new city by interviewing dozens of her human neighbors. A retired bodybuilder who explained how the expansion of the highway through their neighborhood impacted his lungs; a widow who'd missed her daily small talk with neighbors that had been driven out by gentrification; a timid immigrant trying to find his place among the cargo bikes brimming with blond toddlers. Soon, Puck realized that their feelings of concern and disorientation were not unlike those of the trees. Between the rising pollution in the air and soil and the impact of disconnecting urban trees' roots from one another, severing their vital nutrient-sharing over the "wood wide web," the trees that had watched a city rise around them must have felt stunted. Puck decided to research their life stories to share them in a new way, a form that could firmly root her neighbors and herself in the place they'd chosen to call home to motivate them to preserve it. Like the Victorians who planted trees and landscaped romantic fields amid a smoke-clogged London, she hoped that giving the neighborhood's residents a means of connecting to the nature they had could offer a salve for their sometimes-challenging lives. The result was "Giants of the North." What you feel empathy for, Puck understood—whether a human, tree, or otherwise—will be what you protect.

Marije, the oak who spoke to me, told the same one that millions of other trees might tell if we gave them the voice to speak. As we have discovered, life on Earth is a gossamer web of fragile relationships that frays exponentially the further we slice into it. As a result of human activity, one million species currently struggle at the brink of extinction. Resource exploitation, pollution, invasive species, and, most critically, habitat loss are all to blame.

As the tree's story reveals, the industrial realities that seem imperative to our way of life amount to a death wish for so many of the organisms that, in the long term, we rely on. Even worse, as the destruction mounts, the species it threatens are growing closer to home.

If the dangers of our behavior could be contained to nature while the progress of human civilization marched on like an assembly line, perhaps some of us wouldn't mind. Though our green spaces and animal kingdoms might suffer, those costs could be justified as the price of feeding and housing our growing population or bringing more humans out from the suffering that is poverty. Yet, as we are beginning to discover, the fate of the natural and human worlds lies inseparable. Our disconnection from the landscapes around us isn't only killing nature, it's beginning to kill *us*.

"I'm not really a people-tree. I'm more of a tree-tree," Marije told me. "I—"

"What are you doing?" a voice behind me cut Marije off. I gasped softly—I'd been so engrossed in the tree's story that I'd forgotten where I was.

I turned around and found a girl, about six years old, looking up at me. "What are you listening to?" she asked, pointing at the oversized headphones on my head.

I pull one of them back to hear her better. "The tree," I said, laughing. "I'm listening to the tree."

"Trees don't taaaalk," the girl snickered.

"Oh yes, they do." I smiled at her. "Try this." I kneeled and put the headphones from my head to hers. Realizing they were too big for her small head, I held them in place.

"I don't hear . . ." she started saying.

"Shhh, give it a second." Her impatience reminded me of my own. I pressed the play button on my phone.

Her eyes bulged. Then she swallowed as if trying to digest the shock and wonder streaking across her face. She was nodding now. "Yes, I can hear you," she whispered.

I let her listen to Marije's whole story, whose ending I hadn't heard yet. Seeing the girl nodding intently almost brought me to tears.

She shrugged the headphones off and handed them to me.

"What did you think?" I ask.

Instead of answering, she ran to Marije, wrapping her little arms around her trunk.

"I'm here, Marije. I'm here."

"Giants of the North" is one of hundreds of experiments launched worldwide in recent years to leverage new technologies to build connections with our natural surroundings. Puck tells me she knows that some might critique it for anthropomorphizing trees. Still, her goal was for people to understand the neighborhood better and increase their empathy for trees so that they're more likely to treat the nature that has survived urbanization with respect.

Part of what made the installation so effective was how completely it absorbed you. Instead of asking you to scan a QR code at each "stop" like most audio tours do, "Giants of the North" uses GPS to trigger the monologues. You put your headphones on, and as soon as you arrive at each "geo-fenced" location throughout the one-hour tour, the respective tree's story begins. There is no tech distracting you from the story, nothing but a passing car—or, in my case, a curious child—to interrupt the magic.

"Ultimately," Puck tells me, "the aim is to change how we relate to these trees around us." I live three miles from these trees, and though I only listened to the audio tour once, I never walked by Marije again without thinking about her story.

The project has shown me firsthand the power of the right nature-driven tech. The uncanny connection its stories built, startling me into paying attention to plants I'd walked past in my neighborhood for years, demonstrated how recent innovations can build empathy with the ecosystems around us—in both those who already love nature and those who don't yet realize it. After I encouraged a spatial planner for the city of Amsterdam to check out the audio tour, he emerged an hour later as a changed man. He is now campaigning to have "Giants of the North" installed as a permanent fixture, and he wants the city to require that all employees experience it as part of their onboarding. For someone who has devoted her career to reworking technology to solve many of the problems it unleashed, such experiences fill me with hope. They remind me how we can regain the solidarity with nature that we have lost in cities designed around automobiles and climate-controlled structures.

Across the world, technologists and designers have launched projects that boost mental health by building relationships between city residents and our natural surroundings. One of them, Yvonne Lynch, draws her inspiration from the story of a woman who had bonded with a tree.

Yvonne, who at the time was the head of the Urban Forest and Ecology team for the city of Melbourne, learned of the story through an email. Its author, Sandra, shared that she had given birth to premature twins in the neonatal intensive care unit of the Children's Hospital in North Melbourne. Sandra wasn't allowed to stay overnight in the hospital; she dreaded having to leave her children every night.

Walking out of the automatic doors, making the long drive home, and waiting through another sleepless night, wondering whether her daughters would survive, were the darkest moments of Sandra's life.

But before she drove home to collect herself, Sandra found the strength to make it by collecting her thoughts under a large gum tree just outside the hospital, admiring its strength. Eucalypts—often called gum trees—are icons of the Australian flora. With over eight hundred species, they dominate the Australian landscape, forming forests, woodland, and shrublands in all environments except the most arid of deserts. They are also hardy and resilient to pests, a testament to their long-fought battle for domination in Australia's backcountry, making them ideal species to plant in cities. I later discovered through Melbourne's tree map that this gum tree is glossy green in the summer. It transitions into beautiful colors in the fall—red, orange, yellow, and even purple. The bark of this variety is silvery and smooth to the touch; at times, the trunk feels almost naked. Probably the most notable feature is its smell, a multifaceted forest scent with hints of mint, honey, and citrus, which some people describe as sharp and slightly medicinal. Whatever it was, the tree provided strength and solace to Sandra when she needed it most.

That was fifteen years before she wrote the email. Now, when Sandra drives past the tree, her teenage twins are in the back.

Yvonne tells me about Sandra while I sit in a coffee shop on the west side of Manhattan. I sit outside, bundled up against the briny wind that sweeps toward me from the Hudson River, a few blocks away. When she shares Sandra's story, her mascara runs down her cheeks like a branch heavy with apple blossoms. Yvonne whispers something so softly I have to hold my laptop to my ear, a confession. The email wasn't addressed to her—but to that gum tree.

"Wait—" I say, wondering if I've misheard. "How do you email a tree?" I ask her to retrace our steps. Yvonne tells me that the email was just one of many eye-opening revelations from a project she began in 2012. At the time, she'd encountered a major problem in her role over-seeing Melbourne's forest: thousands of trees across the city were dying from drought. Melburnians were understandably frustrated and had many questions—which Yvonne had difficulty answering. "This branch is going to come down on my car ANY second!" wrote one homeowner in the first of many alarmed messages overflowing from her email and voicemail. "I'm really worried about the color of the leaves," another ag-itated constituent wrote.

Keenly aware of the limits of email when emotions ran hot, Yvonne decided to engage these concerned citizens over the phone.

"Will you plant a new tree to replace the one that died?" asked a small voice on the other end of the line.

"Which tree do you mean?" Yvonne replied.

The questions were never-ending. Yvonne couldn't keep up. Frus-trated, not only about losing so many trees on her watch but also because she couldn't help people fast enough, the stress was getting to her.

So she designed a modest experiment in a last-ditch effort to help Melburnians. The idea was simple: assign each one of the city's 70,000 trees an ID number and an email address. The system would sidestep the endless back-and-forth so often required to pinpoint the problems citizens reported. At the time, Yvonne expected the tree labels to make her office more efficient. It was a humble, if respected, program that many thought wouldn't amount to much.

What happened next took everyone off guard. People started to write love letters—not to Yvonne or her staff, but to their favorite trees. Thousands of emails started coming in weekly. "Dear 1037148," one admirer wrote to a golden elm. "You deserve to be known by more than a number. I love you. Always and forever."

"You are beautiful," wrote another. "Sometimes I sit or walk under you and feel happier. I love how the light looks through your leaves and how your branches come down so low and wide it is almost as if you are trying to hug me. It is nice to have you so close; I should try to visit more often."

Inevitably, the experiment also brought out our human foibles. Some confessed to tree adultery: "I have fallen in love with 1583182," one local wrote, seeking advice. "I feel guilty for cheating on 1023379, but I justified it because 1023379 is expected to die in under a year." The heartbreaker wanted advice: "Should I wait for 1023379 to die, or should I leave 1023379 for 1583182?"

Sifting through the tens of thousands of emails—and driving the streets of her city to spot the very specimens that inspired these emotions—reminded Yvonne why she had thrown herself into the gauntlet of public service in the first place. It transported her from the brink of burnout to a new, expansive sense of possibility. In a year in which Yvonne had seen so many trees die, starting to question the importance of her work, the emails inspired her to keep at it. "I had always prided myself on forging policy pathways for nature without getting soft," said Yvonne, explaining that those who show emotions for trees often get dismissed as "tree huggers." Revolutionizing how residents of her city shared their innate love for nature allowed Yvonne and the city's other residents to see that the emotions were not something to run away from: by embracing our inner tree hugger, we could all be more grounded.

The emotional outpourings to trees, such as the gum tree outside Melbourne's Children's Hospital, were unintended, though by no means unwelcome. Eventually, after the project went viral, people worldwide began to show Yvonne what a tree on a sleepy street in Melbourne meant to them. "Nature can be there for you in those difficult moments when people cannot," Yvonne said. "That's one reason it's vital for mental health and an understanding that needs to be reflected in every city's greening policy."

Six years after the launch of Yvonne's Email-A-Tree program, a young student in Halifax, Canada, was so inspired by the organic engagement of the initiative that she decided to dedicate her master's degree to furthering it. The student, Julietta Sorensen Kass, wanted to connect people with plants to address the highly prevalent malaise we often feel after a week in the office. She imagined she could do that by making interactions with the trees more conversational, which she accomplished by moving the interactions from email to text and enrolling a contingent of volunteer "tree speakers." These volunteers created personalities for each tree and agreed to answer questions, typically leading to long, meaningful discussions. Communications were managed through a web-based customer-service platform called Zendesk.

If you could talk to a tree, Julietta had wondered, what would you say? When she put the campaign into motion, the public answered.

Over two months, nearly three thousand people sent over ten thousand texts to fifteen different trees. Many texts were appreciative: "You have been here since I was little and used to come with my grandparents. We always got ice cream after enjoying the gardens. Thanks for your shade and beauty!"

Others sounded apologetic: "Thank you for your service, dear tree!

Do you feel humans are grateful toward you? We hit you with a Frisbee today. Sorry about that."

Some were inquisitive: "My daughter wants to know: do you have a baby tree?"

Those that surprised Julietta the most—the ones that her "tree speakers" forwarded to her daily—were downright heartbreaking: "Hello Miss Luna Ruby," one began, using the name Julietta had given to a weeping beech tree, "I am so grateful to have this opportunity to tell you the comfort you have provided. You may not know even though I have sat under your leafy canopy, but your heavy, gnarled branches that exude wisdom were there during my most difficult times . . . the death of my mum, my dad, and even during a frightening diagnosis. You may not have been able to reach out with your branches, but you provided comfort when I needed it most."

These texts gave people a means of exploring their feelings for nature—and the surprising impacts that nature had on their emotions. I walk through some of Julietta's favorite urban forest patches in the Halifax neighborhood where she used to live. She tells me that she hoped the exchanges might help people step out of a difficult headspace for a few minutes and redefine what they were going through. "Most of us spend so many hours every day staring at a screen," she explains, "working through abstract problems that our ancestors could have never imagined."

Leaving the office, you are typically full of all anxieties and emotions you may not know how to process. It gives you perspective, she adds, when instead of walking right past a tree—an organism that has been there for decades, or sometimes even centuries—you can connect with it. These plants remind us that life moves forward, challenges fade, and new ones emerge. The way great things are built—like these towering branches above us—is through persistence.

Over the phone, Julietta leads me around the corner of a brick building, and we begin to walk down a boulevard with wide-trunked ash trees rising from the center, separating traffic. I asked Julietta which text message touched her the most. She pauses, finding the words. "There was one participant," she continues, "whose grace and resilience I will never forget."

An air horn blares its two-tone note as a truck speeds through the crosswalk before me, just a few steps away, casting a cloud of exhaust. I'd gotten so absorbed in understanding Julietta's mission that I'd walked into the street before the signal turned. "My fault," I say, trying to steady my nerves.

Julietta tells me not to worry. "That way," she says, and I see a pocket park where a semicircle of well-manicured shrubs forms a wall around a few small benches surrounding concrete chess tables.

"The participant," I remind Julietta, taking a seat. "Why was that person someone you will always remember?"

"What she shared was so charged," she says, "so emotionally challenging. I wanted to invest great care and respect in crafting the response."

I asked her what the text was, trying not to let my curiosity sound impatient. Julietta pulls it up: "It's been a bit of a rough year," she reads aloud. "My husband and I have been trying to have a baby for a few years . . . I just lost my baby," Julietta tells me. The woman went on to describe the complications of her pregnancy, the challenges of accepting and coping with the loss, and the hope that in the future, she would be able to bring her children to meet the tree. From these texts, Julietta could tell that the messages were meant for the tree, not the human volunteer tasked with giving the tree a voice.

Text-A-Tree is a prime example of how technology can enhance our connection with nature so that we can draw emotional resilience from our surroundings. Giving people a means to communicate with urban nature using the tools that dominate modern life makes the natural world accessible. It rekindles the relationship at the core of the hunter-gatherer and agricultural lifestyles that supported humans for most of our existence, allowing us to understand ourselves better. While more research is needed to conclusively demonstrate why it is that exposure to nature proves so vital to mental health, there is no doubt that programs like this one make it easier to connect with our environment. Or that the cities that incorporate nature-oriented activities into their residents' lives will unlock vast benefits for those people's emotional well-being.

Since completing the project in Halifax, a lot has changed in Julietta's life. She moved back to her hometown to be closer to family; founded her own business, which aims to bring more Text-A-Tree-like experiences to urban dwellers worldwide; and has become the mother of a baby girl. However, not a day goes by when she doesn't think about the texting trees, because it continues to remind her of the healing possible when we foster human-nature relationships.

"Technology has become such a constant part of life in Western society," she says, "that it now informs how we experience the world. Suppose we want to transcend technology and bring ourselves closer to our ancestors' relationships with nature. In that case, we need to use engagement models that align with today's world—models in which technology can promote relationships between people and nature. The most effective method will likely emulate how people develop and build relationships with one another, and for many people worldwide, that method is texting."

I sit for several minutes without speaking, watching the shafts of the evening sun—filtered through the bushes' leaves—make ornate shadows on the concrete table before me. I can feel my breathing ease and my blood pressure drop.

Text-A-Tree was the world's first, albeit still unpolished, attempt at exploring the role of texting in fostering human-nature relationships. Not only did the engagement far exceed expectations, but members of the public clearly engaged with urban trees and, through their interactions, appeared to develop relationships with individual trees. While there seems to be little question that Text-A-Tree was a success, Julietta admits that she sees it more as one minuscule proof of concept that illustrates all the human-nature interactions we should pursue. The project showed that many people are comfortable forming relationships via text with trees, she explains. And given that, there's so much more work that cities and institutions could be doing to foster these relationships: gardens and arboretums could launch a phone number to connect visitors to plants; cities and towns could create automated treasure hunts where clues are gathered by texting plants in the community; municipalities could develop interactive walking trails in which textable trees offer directions, encouragement, or information; hospitals, rehabilitation centers, and mental health networks could launch "Grandmother trees" to provide asylums in which participants can share feelings and experiences, and seek help from trained personnel; schools could offer counseling services mediated through schoolyard trees.

"There's one more tree I want to show you," Julietta says, and I am off. She warns me that it may not look like much, but it is her favorite natural landmark near her old home—one that orients her in her life and gives her strength when she sees it every morning. Her giddiness at the everyday wildlife in her virtual backyard reminds me of my spiritual reawakening to the nature around my childhood home during the pandemic.

"When everything is stripped away," Julietta tells me, I am looking up at what I now understand is the tree that inspires her, "we cling to life—to pets, to plants, and each other." Ultimately, Text-a-Tree allowed people to recognize trees as individuals—even as valued community members. Not only for how much carbon they sequester or how well they filter our air, but perhaps even more importantly, for their social significance as institutions whose endurance inspires us. When all our best-laid plans seem to fail us, nature's marvels remain.

Experiments as seemingly bizarre as texting a tree further the field of ecotherapy, and they are quickly gaining acceptance in the scientific community. As ecotherapy gathers steam, some doctors have begun to write prescriptions you cannot fill at the pharmacy. Instead of pills, "nature prescriptions" can help patients cope with various conditions, including chronic illnesses such as diabetes and high blood pressure—and most notably, mental health issues that have surged in recent decades, like anxiety, depression, and attention deficit disorder.

Most of us don't need a prescription to understand why nature can profoundly impact our mental health. We've all felt it at one time or another: the calm that comes over us when surrounded by greenery. Or the cheerfulness we feel when strolling in our flip-flops, we first see beachgrass rising from the dunes and can hear the crunching sea. But in the last two decades, scientific research proving the benefits of spending time in nature has grown exponentially. Emotional experiences as important to a healthy outlook as finding peace, being able to let go of things, thinking more clearly, thinking more positively, worrying less, feeling hopeful, and feeling powerful are all correlated with spending more time in nature, says Dr. Lotte Mortier De Borger, a family physician.

For her master's thesis at the University of Antwerp, Lotte prescribed time in nature to her patients. "It's never the only solution," she's quick to disclaim, "but it's an essential strategy for treating mental disorders like anxiety, depression, and burnout." Even in her own life, Lotte believes spending time in nature is vital to maintaining a healthy lifestyle.

The mechanism behind exactly why nature has such profound impacts on mental health is more difficult to grasp. There are three prevailing theories. One provides that nature immediately evokes a strong positive feeling because humanity has an evolutionary predisposition to respond positively to nonthreatening nature. This explanation follows from the insights of E. O. Wilson, the legendary ecologist who dedicated his life to conserving nature. In his "biophilia hypothesis," Wilson suggests that humans are genetically programmed to be affiliated with the natural world.

Another theory states that nature demands attention without effort, allowing the brain to recharge. Known as Attention Restoration Theory, or ART, it proposes that exposure to nature is enjoyable and can also help us improve our focus and ability to concentrate. This hypothesis was developed and popularized by Stephen and Rachel Kaplan in the late 1980s and early 1990s, characterized by rapid technological advancement and ever-increasing indoor entertainment. In a 1993 study involving participants in what is arguably the most stressful time of their lives—when they recovered from breast cancer—a researcher found that cancer patients who spent time in natural, therapeutic environments showed improved performance in attention-related tasks, higher likelihood to not only go back to work but also return to working full-time, greater inclination toward starting new projects, and higher gains in quality of life.

Another explanation is the Stress Reduction Theory (SRT) coined by Roger Ulrich in 1981, which proposes that natural environments promote recovery from stress, while urban environments tend to hinder the same process. Sonja Sudimac, a young researcher in Berlin, tested this theory. A Ph.D. candidate at the world-renowned Max Planck Institute for Human Development, Sonja had been fascinated by how the environment shapes our brain, specifically how a natural environment can influence stress and emotions. Sonja knew previous studies had shown that the amygdala, the part of the brain that controls emotions, such as fear, sadness, and aggression, and prepares the so-called fight-or-flight responses to danger, was more activated in urban residents than in rural dwellers. However, no study has examined the causal effects of natural and urban environments on stress-related brain mechanisms to make sense of why this difference in response existed. Perhaps most importantly, no previous findings disentangled whether stress relief after being in nature results from exposure to the natural environment itself *or* merely the absence of detrimental urban effects. So Sonja got to work designing a study in which sixty-three participants went on a one-hour walk in either an urban (busy street) or a natural environment (urban forest)—and measured brain activation before and after the walk.

"We had hypothesized that nature would decrease amygdala activity, while a walk on a busy street would increase it," Sonja tells me. "But it surprised us that walking in the city did not cause additional stress-related brain activity." Instead, she found that the brain activity in stress-related regions remained stable after the urban walk, which suggests that the stresses inherent in urban life may be canceled out by outdoor exercise—even if that exercise is on city streets. Sonja posits that spending more time in nature might increase the amygdala's threshold for activation. In other words, nature exposure could build a buffer against the stresses inherent in urban living and lower the risk of mental disorders among city dwellers.

Ultimately, biophilia, ART, and SRT are deeply complementary theories: all three help to explain the mechanisms driving nature's profound impact on the mind. We humans are both innately complex and wild beings. Our ancestors evolved in wild settings, relied on the environment for survival, and were driven to connect with nature. There is abundant evidence that the same urges play as much of a role today as they did 200,000 years ago. Because of those urges, nature can replenish our cognitive resources, restoring the ability to concentrate and triggering a physiological response that lowers stress levels. Furthermore, depriving oneself of nature can kick off a vicious cycle, requiring more and more time in nature to trigger these physiological responses and lower stress levels.

The next step, Sonja says, is to figure out exactly what aspects of nature are reducing amygdala activity when one takes a walk outside. "Pinpointing the specific features, like colors, sounds, odors, access, diversity, terpenes"—which I'd learn are the naturally occurring chemical compounds that make plants smell the way they do—"Could inform how we design our cities." The data could improve how doctors prescribe nature-based therapies, how urban planners build green spaces, and even how architects design buildings to keep office workers in touch with natural stimuli such as light and fresh air. Together these tweaks could protect and improve residents' mental health.

One major challenge doctors encounter when encouraging patients to increase nature exposure is nature's amorphous, subjective qualities. Nature has a very wide definition, and what constitutes relaxing wildness to some may not to others. Nature can mean green spaces like parks, woodlands, or forests, and blue spaces like rivers, wetlands, beaches, and canals. For those most deprived of it, it might also include trees on a street, private gardens, road verges, and indoor plants or window boxes. Even watching nature documentaries can improve our mental health. These varying understandings of nature are good because they make some of it accessible to everyone. But as some therapists have found, they also make prescribing the right exposure to nature—and studying its effects—difficult.

To make it easier to prescribe nature, Dr. Robert Zarr, a pediatrician who was an early proponent of nature prescriptions, founded Park Rx America, a nonprofit database of nearly 10,000 parks, which physicians can search based on dog-friendliness, wheelchair access, and parking options. When I connect with Robert over Google Meet, his head bounces up and down on my small screen. I discover that, true to form, he's taking our interview outside, getting in his steps in his new hometown of Ottawa. At some point, his arm tires of holding it out in front of him, positioned on his bouncing head, so we turn the cameras off. Robert tells me that the filter feature on Park Rx allows practitioners to prescribe a park that fits each patient's unique needs. Since its conception in 2017, over a thousand registered healthcare professionals have incorporated the tool into their treatment plans.

Jocelyn was a young recipient of one of these prescriptions. Robert met her in the emergency room in Washington, D.C., where she complained of dizziness and debilitating chest pains. For an otherwise healthy seventeen-year-old, the symptoms were troubling. After a thorough investigation and many tests, Robert concluded that there was only one thing clinically wrong with Jocelyn: she was stressed. The noise of all the chatter in Jocelyn's life, all the pressures, all the stresses, and all the expectations had taken a toll on her well-being.

Where another doctor might have told Jocelyn how to relieve her stress, Robert had learned that the most effective plan is the one patients will drive themselves to maintain. So, dead set on ensuring he would never see Jocelyn in the emergency room again, Robert asked her what she liked to do outside. As it turns out, no one had ever asked Jocelyn that question, and her answer surprised him: "I have a hammock in my father's backyard," she said. "I like to lay there."

That became the basis for Jocelyn's nature prescription.

"A prescription," Robert tells me, "has four important components: the name of the medication, how many, in what frequency, and how long." I hear his even phrases accented by his rhythmic steps, his shoes flexing on what sounds like gravel. "I've adapted this conventional prescription for medicating using nature," he adds. Jocelyn's prescription was for her to lie in her hammock and observe the clouds above her—and the wind rustling through the branches of the trees—for sixty minutes once a week. The most important thing was that Jocelyn immersed herself in the experience, not lying in the hammock dozing off or mindlessly scrolling through her phone.

At first, Jocelyn was bored out of her mind. More than once, she called Robert and asked him if she could lie in her bed instead. She felt silly, desperately waiting for her alarm to go off, signaling the end of her weekly session. But after a few weeks, something shifted. For the first time in her life, she noticed the leaves change colors, the birds sing, and the wind whistling. Jocelyn had never seen the clouds shapeshift and dance across the sky. She abstractly understood that they did, but they seemed still every time she looked up. Now, she saw clouds drift and transform in front of her. Soon, she became fiercely protective of her Wednesday afternoon ritual in the hammock, sometimes going outside even in the rain. "Doctor's orders," she'd say if her family tried to redirect her.

After a couple of months, her chest pains went away, and she steadily had fewer and fewer headaches. But the effects of the prescription, something she continues with today, are much longer lasting. "The old Jocelyn," Robert recalls, "was quick to snap at just about anything and everything. She was easily agitated, annoyed, and even violent. The new Jocelyn is a problem solver who keeps her attitude in check. Now she even shares the wonders of nature with her much younger siblings, hoping that when life becomes difficult—as it inevitably does—they'll find solace in the outdoors."

By prescribing nature using the Park Rx America platform, clinicians *and* researchers can take advantage of the added value of two break-throughs. First, highly specific prescriptions, which account for factors including the place, activity, and frequency—like one-hour weekly ham-mock lounging and observation sessions—to further study nature pre-scriptions' impact on health. And second, electronic prompts take the first step to redirect patients' trajectories of malaise into ones of health, reminding them of their "medication" when they pass outdoor spaces that satisfy the characteristics of their prescriptions.

It's no coincidence that so many of big pharma's advertisements bom-barding your TV—families hiking through mossy woods, retirees play-ing in the surf, or couples lounging in outdoor baths before a dramatic vista—have something in common. The actors making the case for the industry's latest miracle formula all seem to be already taking the orig-inal panacea: nature itself. The next frontier of technologies support-ing our mental health may make this miraculous engine of health a tool available for everyone.

My bike slows to a halt across the street from the sprawling oak. It's been three weeks since I listened to Marije's story, feeling her pain and imag-ining her suffering. I see a boy, not older than seven or eight, climbing Marije, scraping his way up her trunk and grasping onto her branches.

I feel my brow furrow and my face winch. You're hurting her, I think. I'm about to cross the street and tell the boy to get off her, to leave her alone, when I swear I hear Marije's voice.

"It's okay," she creaks, "this is good. We came from the trees—this boy knows that."

Marije's right. Some scientists speculate that since humans evolved with trees, using them for food and protection, our affinity for them is likely programmed into our DNA. We are hardwired to love trees. For the protection they offer us from wind, the shade on hot summer days, the wildlife they attract, and the food and lumber they provide us with.

But trees offer more than provisions.

We carry strong memories and emotional ties to people, places, and events through trees. We love trees like we love our friends and family. For some of us, they *are* friends and family. Modern urban life has anchored us to an indoor existence and abstracted us from the natural environments we evolved in, disconnecting us from the elements that support our survival. Growing up isolated from nature can have an even more catastrophic effect: at its worst, it can make us afraid of nature— nature-phobic—triggering a destructive cycle that leads to total nature deprivation and all the negative emotional consequences that follow.

Combating this fear—learning to love what modern life has rendered a perilous mystery—takes time. Love, as we know from our human relationships, requires work. But as Jocelyn learned, if we want optimal mental health, we must make the time and space to tend to our most important relationships, especially our connection to our habitat. Immersive experiences like "Giants of the North" offer a bridge to get us there, transporting us from a world locked into screens to one where our devices lead us to explore the fascinating life around us. After we recognize our illness—the imbalance between nature and technology in our urban lives—the first step is to abandon the value-laden dichotomy of nature *versus* technology. Used responsibly, technological breakthroughs—the thing that once drove us away from nature—hold the secret to restoring what we lost.

CHAPTER 8

Into the Great Nearby

Nature itself is the best physician.
—Hippocrates

THE BRANCHES TAPPING THE small window of the attic were like outstretched arms, beckoning for an embrace. Through this window, fourteen-year-old Suzy would escape, seeking refuge in the then seventy-year-old Linden tree. Suzy was from Amsterdam but spent three long years hiding in the attic of a house in Eefde, a village in the Dutch province of Gelderland, twenty miles from the German border. It was 1943, and Suzy was Jewish.

Amid the branches of the towering tree, Suzy found solace, indulging in books, keeping up her studies, and coming to grips with the atrocity of living in hiding in the Nazi-occupied Netherlands. Veiled by the cloak of lush foliage, she eluded the eyes of the German military. In the precious hours she spent high up in the tree, she was free.

More than just a physical shield from danger, the tree provided Suzy with a sanctuary—a place of peace and solace amid the chaos of World War II. The tree kept Suzy's life manageable in many ways, its branches serving as a lifeline. For this Holocaust survivor, the linden represented hope, resilience, and the triumph of nature amid the darkest circumstances, a testament to the power of the natural world to heal, protect, and sustain us—even in the face of unimaginable adversity.

The tree is now 150 years old, having lived through innumerable joys and unimaginable cruelties. For those who know its history, trees like this offer a bridge to generations who came and went, savoring its shade and silent beauty. They also represent a connection to the children of the future who will grow old in this landscape after we are gone.

I learned about Suzy and her lucky Linden through a contest of sorts. In September 2022, I was thrilled to get a call from two people I had come to admire. One of the voices on the other end of the line was Dirk Doornenbal, managing director of De Nationale Bomenbank, or the National Tree Bank, a leading tree-care company in the Netherlands. The other was Henry Kuppen, managing director of Terra Nostra, a research institution for trees and soil. Dirk and Henry had teamed up to celebrate De Nationale Bomenbank's fiftieth anniversary in a unique way: by asking the public to nominate their favorite trees.

These nominations (ultimately ninety-eight were submitted) would be narrowed down to the top fifty. Dirk and Henry wanted me and three other "professional tree hugger" jury members to select the top five. These top five gentle giants would not only get the attention they deserved but, more importantly, lifelong tree care to ensure they would be there for generations to come.

This project was deeply personal for De Nationale Bomenbank. The organization has made its name in the competitive tree-care space due to its workers' talent and skill in preserving, transplanting, and often saving large, centuries-old trees. More than almost anyone else, these committed tree caretakers understand the critical importance of preserving what we have rather than focusing primarily on planting new trees.

After we selected the top five, each jury member could choose one of the trees to act as its "ambassador." I chose Lucky Charm, the now 150-year-old linden that kept Suzy safe. According to the tree's current "owner," Ton Rotteveel, Suzy found peace in the chaos under the cloak of the tree's crown. Ton insists the tree continues to bring him and his family good luck.

The human mind has a peculiar connection with nature that almost defies explanation. It may be the way the lush greenery sways gently in the breeze or how the sound of a babbling brook mimics the rhythm of our heartbeat. Whatever it is, there's no denying that nature has a calming effect on our minds, allowing us to escape the chaos of daily life, even if just for a moment. Nature has proven, time and time again, to be one of the most effective tools in improving mental health, used as a remedy to soothe our minds when we're feeling overwhelmed or stressed out. Its calming and therapeutic effects have been well documented, but recent scientific research reveals that the benefits of nature extend far beyond easing our mood. Spending time in nature helps lengthen our life span.

How is it possible that something so widespread could hold the secrets to longevity? One theory is that when we spend time in nature, our bodies release feel-good chemicals that reduce stress and boost our immune system, lowering our risk of developing chronic illnesses like heart disease and cancer. Additionally, nature provides fresh air and natural sunlight, which positively impact our bodies. Immersing in nature helps us prioritize self-care and achieve balance, providing opportunities for physical activity, leading to a longer, healthier life.

The uplifting presence of Lucky Charm speaks to a greater truth. We are irrevocably connected to the natural world. In our pursuit of progress and modernity, it's become all too easy to ignore how the earth sustains us. But our ignorance makes the power of nature no less important when fighting off stress and anxiety and lowering the risk of chronic disease. Scientists now attempting to identify the mechanisms through which nature sustains us recognize the power that Suzy may have experienced in her precious time outside—the key to human longevity.

Too often, cities demonstrate the flip side—how an absence of nature challenges our ability to persevere. The thought might strike you as you walk to work, your eyes stinging from air thick with smog, or as an eighteen-wheeler blows past you on your bicycle nearly knocking you to the pavement, driving home just how hostile your environment is to your presence.

A few years ago, at a drugstore in Toronto, I was looking for an ointment to treat a cut I got slipping on a poorly drained sidewalk, where the thawing snow had refrozen into ice. A face mask caught my eye in the skin care aisle: Nivea's Urban Skin Detox. Specially formulated for "city girls" like me, the facial care line promises to protect against uniquely urban threats like air pollution. After thorough cleansing and treatment, the product claimed that my skin would emerge "#cityproof." The face mask would function like armor to be put on before battle.

I was furious. Not at Nivea, who arguably did what any corporation would do in addressing a growing and potentially profitable problem— but at urbanists like myself whose collective decisions had transported us here. How had we let city air get so bad that we needed a shield to protect ourselves?

From that moment on, I looked at my hometown differently. It made me reconsider every thread in its urban fabric. The fast-food joints are replicated on every corner. The local bodegas lacked a fridge for produce. The parents who would rather play taxi driver for decades than send their kids to school on a bike. Even the police charging jaywalking as a misdemeanor, rather than seeing it as a desperate plea for a crosswalk, couldn't avoid my ire.

It is no wonder that, as the nation's cities have swelled over the past decade, U.S. life expectancy has stagnated. Despite the trillions of dollars Americans spend annually on healthcare, public health has not improved. Instead, the prevalence of chronic diseases like heart disease, cancer, strokes, Alzheimer's, and diabetes are soaring, and most Americans are barely moving enough to meet the surgeon general's already dismal physical activity guidelines.

Instead of adapting our environments to our health needs, society has put the blame squarely on the shoulders of individuals. For decades, the public has shamed the obese, diabetics, and sometimes even the cancer-ridden, suggesting they have caused their own suffering. Even though we all must answer for our choices, leaders are too quick to adopt the rhetoric of personal responsibility simply to dodge their complicity in our ballooning health crisis.

Anyone who wants to make healthy choices in America today knows how difficult it is. So many of us live in neighborhoods without green spaces to play in, without public transit or cycling infrastructure, or where temptations lurk in vending machines in every corner of our offices and schools. And suppose we manage to find the time to exercise in our stressful lives. In that case, the limitations of our outdoor spaces push us onto stationary bikes, where we either stare at news headlines that spike our cortisol levels or "travel" through virtual landscapes from within a windowless basement gym.

The fact that we so often fail to follow the health guidelines of our leaders is not our fault. The environment surrounding us is unmistakable in shaping our lived experience. It can constrain our behaviors and lifestyles, which in turn limits our physical well-being. Similarly, it can facilitate behaviors that improve our longevity beyond what we thought was possible. Only by redesigning the built environment and altering its structure can we empower ourselves to change the perilous trajectories of our physical health.

For nearly two hundred years, ever since John Snow mapped the victims of large cholera outbreaks in London, we've known that where we live, work, and play strongly influences our health. Now, as the mounting failures of urban design overwhelm centuries of progress in medical care, the relationship between city planning and physical health is unavoidable. We shouldn't be content to tread water or move backward as doctors strive to offset the consequences of cities that slowly kill us. Instead, we should demand that our environments provide the lifestyle we need to further our longevity.

Furthermore, no amount of medical interventions is likely enough without addressing the pitfalls of our modern habitats: the common denominator across cities where people live long and happy lives has nothing to do with hospital access or medical schools—it is access to nature.

Jared Hanley separated from his team on day eight of a nine-day adventure race. His three teammates had just rappelled down a steep Montana cliffside on their last stretch toward victory, carrying all his camping gear. It was Jared's turn to rappel when lightning struck.

"Rule number one, in the event of lightning," Jared tells me, relaying the terrifying moments of his ordeal, "is to get the hell off the mountain." But when he reached the cliff's edge, feeling a few of the slick handholds, he saw that rappelling down the wet mountain using his metal carabiner in a thunderstorm would be even more dangerous than the alternative. So instead, he scurried up the cliff and, with no equipment, dug a hole to protect himself. The staticky rain and booming thunder sounded like a lo-fi recording of a drumline completely out of formation.

"There's something about lying in a hole, curled up in the fetal position, under a bunch of branches, soaking wet, chilled to the bone," he adds. "It puts things into perspective."

Jared and I are about to experience a taste of that exhilaration. He'd invited me and his two co-founders to hike Oregon's Clear Lake. We plan to finish the long day with a plunge in the placid spring-fed waters. While it might sound nice, I was anxious. The average temperature of the lake hangs between 35 and 40°F. The lake is so frigid that when it was formed three thousand years ago by lava flows that left behind a basin, the forests submerged within the lake that quickly flooded the terrain were frozen in time. The result is a "petrified forest," remaining today in trunks extending as far as 120 feet into the water, many visible from the surface. I am nervous I won't last ten seconds in the water, though I'm not about to let it show in front of these rugged mountaineers.

Jared was born on the Navajo Reservation to Peace Corps parents who'd just returned from serving in Honduras. By the time they left, at age three, Jared had started to pick up a few words of Navajo, one of the most difficult languages to learn. The language is so complex that the Navajo code talkers used it as the basis for a secret communication system during World War II. Jared's family settled in rural Washington. Shortly after, adventure racing was born in the 1980s, as Jared was learning to kayak, ski, mountaineer, and mountain bike with his father. When Jared learned several years later that he could fuse all these activities that he loved and more into a ten-day checkpoint-to-checkpoint race, he was sold. He earned his stripes in national adventure races. Soon enough, he got sponsored and launched a fifteen-year career competing professionally worldwide.

Through much of his racing experience, Jared kept a regular office job as a data scientist for an investment bank, unsure if he could maintain a living from this passion. He used his vacations to compete. When he returned, ten pounds having melted off his already thin frame, his colleagues would typically welcome him back to the office with a "What the hell happened to you?" He always found his colleagues' surprise at the bruises and scratches he'd earned an amusing epilogue. Jared may have looked emaciated as his body recovered from the crème de la crème of grueling physical endurance, but he *felt* anything but deprived.

"When I spend time in nature, even just for a walk around my local parks," he tells me as we drive through Oregon's countryside, "I'm completely rejuvenated."

Nature, he explains, is his reset button. It gives him physical and mental clarity that he cannot achieve through other means. I laughed, thinking about Jared's co-workers, troubled by his progressively serious hobby when he was undoubtedly in better shape than any of them.

From rural Washington, Jared had moved to Connecticut to study cognitive science, mathematics, and economics at Yale. Then—in his words—he sold his soul to the devil, working at Lehman Brothers during the dot-com boom, taking public big tech companies like PayPal, eBay, and pets.com. Jared loved learning about different companies' technology and understanding the world through numbers, but he found himself repulsed by the culture on Wall Street. He couldn't avoid noticing how he and his colleagues abstracted the concept of "value" from any relation to social good. And the better he got at it, the more disgusted he grew with himself.

After moving to Los Angeles, where Jared didn't fare much better, his wife, whom he'd been with since college, was offered a job as a law professor at the University of Oregon. So Eugene it was. The first time he'd sprinted up the muddy paths between the Douglas firs that lined the city's steep hills, he knew he had found the home he had been looking for. His only problem was that he had no idea how he would help to support his family.

For a time, Jared used his finance tech skills to build employee stock ownership plans, which help employees share in the rewards when their employer succeeds. He even wrote the go-to handbook on this corporate finance scheme. Though he didn't harbor guilt about his job on Wall Street, he didn't find the work particularly exciting. Then one day he watched his former employer, Lehman Brothers, go belly up, setting off the worst financial crisis in almost a century. Millions lost their livelihoods and homes within months, with the unemployment rate soaring to heights that hadn't been seen in generations. Jared decided that year that when he could find the right opportunity, he would pursue a path completely different from any he had ever traveled, something that would help others achieve the peace that he found deep in the forest on a trail race.

We turn up a winding road toward the gateway to Central Oregon's Cascades, my ears popping from the changing altitude. I open my window to breathe in the cool air and shudder when I see the drop-off just a few feet beside the road: a wide canyon of forested hills. Across it, the horizon is segmented by low summits dominated by several soaring volcanic cones.

"So this is where you came to do your thinking?" I ask Jared. Somehow, I already knew the answer.

He offers a tight nod; his eyes trained on the curving road.

Clear Lake is nestled within the confines of the Cascades in the

Willamette National Forest. *Travel + Leisure*, a magazine specializing in jaw-dropping photographs of America's most beloved destinations, had named it one of the nation's clearest, coldest, and most beautiful lakes. Still, even their photos hadn't done it justice. As we turn off the road and into the clearing where a few cars are parked, I'm overwhelmed by the vista. The spring-fed waters are Caribbean blue, the lake crowded by colossal firs, thin silhouettes against a pink sun. A mossy carpet blankets the start of the trailhead.

I crouch down on the moss, feeling its moist and even surface, racking my brain, wondering what it is. I resist pulling out my phone to check.

This was the trail I was running on, Jared says, sitting down next to me, when I had the idea that would change my life.

"So," I said, "what was it?"

"I was circling the lake on this trail, listening to a podcast about Bitcoin for some reason—when it hit me. This system of Bitcoin, and other cryptocurrencies, have no intrinsic value, yet they're worth hundreds of billions of dollars. And here I am, among all these trees, rivers, and lakes that provide untold joy to tens of thousands of people every year. Yet they have no value assigned to them."

He picks up a rock and begins playing with it, cleaning off the dirt. "Some people think it's reductive to try to bring nature into our capitalist domain," he adds, "to reduce irreplaceable beauty to a metric, and I understand that. That was probably why it took me so long to have this realization. But jogging down the trail, which was covered in mist, watching the sun come up over the lake, I finally understood that when it comes to supporting this nature—when it comes to investing in it and preserving it—how we'd declined to value it was exactly the problem."

As Jared understood, giving something a meaningful number allows people to reconcile it with the other features of their lives, and they start using it in their daily calculus. Numbers give definitions to our complex habits and desires that otherwise defy meaning. Valuing a single tree in the abstract is not enough; instead, we must teach ourselves how to appreciate the extensive benefits that nature provides and ultimately use those calculations to improve our lives. This leverages the principles of economics to further our engagement with the natural world, a domain that has previously only been valued when it comes to exploiting natural resources.

Even after conceiving the idea, it took Jared nearly three years to solve the riddle of how to accomplish it. Nature proved to be his toughest puzzle yet, forcing him to quit his well-paid job in finance and coax two people with complementary skills to join him on his impossible quest. The first was Chris Bailey, his friend and former HotelTonight co-founder, who Jared had to persuade to leave behind the early retirement he was enjoying, courtesy of his lucrative sale to Airbnb. The second was Chris Minson, the head of the University of Oregon's Department of Environmental Physiology, who had joined us on the car ride. Somehow, Jared talked these two—whom I'll call Bailey and Minson for simplicity's sake—into joining him in founding a company to solve our value conundrum. The result was NatureQuant, whose mandate is in the name: to quantify nature.

Minson, who had sat beside me on the moss to listen intently to Jared revisit the moment their company was born, suggests we start our hike. Just a few dozen yards down the trail, we find Bailey waving at us from the lake's long dock. He'd driven in from Bend, where he moved three years earlier to be closer to the outdoors.

"Eugene was pretty good," Bailey tells me later, "but Bend is even better." His biggest issue living there is that he can never decide how he wants to spend his weekends outside—whether to hike, bike, climb, swim, paddle, ski, or snowshoe.

As the co-founders discovered, this dilemma was not unique to them—or even to their neighbors in Oregon, graced with such varied landscapes. The deeper they got into the company, the more they realized that they—people who, like me, thought of themselves as "nature-verts" who recharge their batteries through exposure to open spaces—weren't the exceptions.

A large and burgeoning body of scientific literature backs up nature's profound influence on our biology and psychology. Over 150 observational studies and a hundred interventional studies, which together tracked over 300 million individuals from twenty countries, examining one hundred unique health outcomes, have demonstrated the integral role that being in nature has on longevity. That exposure not only has the mental health benefits that Jocelyn experienced when she was prescribed nature in her hammock—lowering blood pressure and reducing stress-hormone levels—but it also promotes physical healing, boosts vitamin D production, bolsters the immune system, and reduces inflammation. While scientists do not know all the mechanisms behind nature exposure, they do know it works. Given this profound relationship, it's essential to understand how much—and of what quality—nature is in your immediate environment.

The research into just how imperative nature is for human longevity is stunning. Decades of studies have backed up the claim that your zip code matters more than your genetic code regarding health. In various cities across America, average life expectancies vary by more than twenty years in adjacent communities—in Chicago, the city with the largest disparity, life expectancies diverge by more than thirty years, depending on the street in which you're living; in both Washington, D.C., and New York City it varies by more than twenty-seven years. But for decades, scientists attempted to explain those disparities purely through social factors, like differences in wealth, diet, social services, and medical care. And while those factors are certainly at play, more recent studies show they miss a key part of the story.

Researchers from Harvard's School of Public Health and Brigham and Women's Hospital, in fact—who partnered with Jared—found, for example, that living in or near green areas can help women live longer. The 2016 study was a nationwide investigation into risk factors for major chronic diseases in more than 100,000 women from 2000 to 2008. After comparing the risk of death with the amount of plant life and vegetation near the women's homes, they found that women living in the greenest areas had a 12 percent lower death rate than women living in the least green areas.

The most compelling finding? As the study's author highlighted, 84 percent of participants live in urban areas, demonstrating that people do not need to move far out to the country to live a long life. The longevity-related benefits of vegetation are incremental, with any increase linked to lower mortality.

The startling impact of nature on health motivated Jared to quantify dosages of this powerful medicine to optimize prescriptions that could increase our longevity and ease the kinds of emotional struggles that Dr. Robert Zarr, the founder of Park Rx America, attempts to alleviate. Building from the NatureQuant technology, Jared designed NatureDose, a personalized prescription app that measures your exposure as elegantly as a pedometer counts steps. The app allows you to set a weekly "nature prescription" goal, like the baseline recommendation of 120 minutes outside, tracking your progress automatically. One psychology professor, Dr. Nick Allen at the University of Oregon, crafted his version of such a tool to help users' mental health. His creation, Vira, uses your smartphone's sensors to collect objective data on your behavior automatically.

In just its first year of use, Vira has saved lives. Growing from research that links suicide rates with nature deprivation, Vira uses its data tracking patient's exposure to nature to send out warnings that doctors and patients have credited with preventing several teen suicides. Considering that the increased rate of suicides have helped to fuel a three-year drop in average U.S. life expectancy since 2014, according to a new study from the Centers for Disease Control and Prevention, we have much to learn.

More comprehensive research proves that nature's miraculous ability to extend our lives cannot be overlooked. A 2019 meta-analysis examined James's results beside those of eight other prominent studies, comparing results from over 8 million subjects. It found that increased greenness around homes is significantly associated with reduced premature death. Each measurable addition of vegetation within 500 meters of someone's home reduced their risk of premature death by 4 percent.

Portlanders are discovering the astounding health benefits of living near trees, as demonstrated in a 2022 study co-sponsored by the U.S. Forest Service, Oregon Health Authority, and local tree-planting nonprofit Friends of Trees. Geoffrey Donovan, a researcher with the U.S. Forest Service, says he started thinking about the relationship between trees and deaths years ago after completing a study that found places with fewer trees also had more deaths. It sparked an intriguing question: Was the opposite also true? Could planting even more trees lead to longer lives? Extensive statistical modeling gave way to a not-so-shocking conclusion: trees planted by Friends of Trees, some 50,000 since 1990, are associated with about fifteen fewer deaths per year in Portland. Geoffrey quickly acknowledged that while his research showed a strong correlation between the number of trees in an area and improved health outcomes, it could not conclusively prove causation. Yet the study's conclusion remains the same even after accounting for race, income, and education disparities among the surveyed population. The results showed Geoffrey that planting trees carries immeasurable rewards, far surpassing the costs involved.

"We estimate the benefit-cost ratio was over a thousand-to-one, which is preposterous," he told a local news outlet. "That's the type of thing you only get in email scams."

When I share these results with the NatureQuant team, they aren't surprised. Not quite ready to dive into the frigid lake, we gathered on a small clearing, taking off our shoes and dipping our toes into the icy-cold water. One by one, Jared, Bailey, and Minson ticked off similarly impressive stats, each riffing off the last perfectly in unison. These guys have done their share of venture capital pitches, I think to myself. But even so, I know they aren't just persuasive—they're right.

"The only issue with these studies," Bailey interjects, "is that they were focused on vegetation, specifically NDVI values." NDVI, or the normalized difference vegetation index, is a simple graphical indicator from a satellite in space to assess whether the observed area contains live green vegetation, like grass, shrubs, and trees. NDVI can also differentiate vegetation from other types of land cover, but anything not vegetation is cast as worthless in the computation's eyes. This is deeply problematic, Bailey explains, because of other health-beneficial natural elements like sand, soil, rock, and water bodies. "Nature is more than just plants."

Minson kicks up a playful splash of water at Bailey. "Are we gonna do this, or what?" he asks. He holds his nose and shoves off the rock with his other hand, sliding into the lake, barely making a ripple.

Bailey continues his thought: "Even lakes may help us live longer," he says with a knowing smile.

Jared nods, saying that was one of the problems they wanted to solve. "If I asked you how much nature is around your house," he adds, "we wanted to give you a meaningful way to account for that."

Minson hasn't returned to the surface as the seconds turn into what feels like a minute. I start to panic, unsure why Jared and Bailey haven't noticed.

"I can't even see where Minson is," I say. "Is he okay?" I jog to the water's edge, and before I leap into the water, I hear Jared chuckling.

He's playing a trick, Jared says. He does this every time we come here.

Minson breaks through the surface about a hundred feet away a few seconds later, gasping for air.

"What!—" I say, suddenly overcome with laughter. "I was about to save your life." Not entirely sure I would've been able to.

"What the hell took you so long?" he shouts back.

I strip down to my underwear. Taking two steps to the lake's edge, I step in. The 40-degree water swallows me, tingling on my skin. I stay submerged for a few seconds, imagining what Minson experienced. Beneath the surface, it's silent but for the cool rush against my ears. My feet find the bottom, a spongy tangle of mud and grasses. For the first time today, I noticed the headache that had been throbbing between my brows—perhaps owing to too much caffeine—beginning to fade away. I return to the surface, floating on my back and trying not to bend or shiver.

I feel a series of wakes as Jared and Bailey join us in the lake.

I focus on taking deep breaths through my nose and out through my mouth. For a minute, my skin burns. Then, I'm overcome with a calm I've never felt before.

Later that evening, huddled around a blissfully searing campfire, crackling and smoking from some of the wet kindling we'd collected, I learned the rest of the story behind NatureQuant. How, after no small number of stumbles, Bailey and a lean team of software developers discovered they could apply computer vision on high-resolution satellite imagery to distinguish all sorts of nonplant nature that had confounded early models, parsing buildings from roads, grass from trees, ponds from the sand. How they programmed the insights about nature-related data that Bailey and other researchers developed into their software, layering the computer-vision's observations with information about parks, light pollution, air quality, tree canopy, and the health outcomes of people in each area. How the resulting software sought to explain changing rates of heart and vascular disease, cancer, chronic respiratory disease, strokes, Alzheimer's and related forms of dementia, and diabetes across the world's natural geography. Jared called the result Nature-Score, a system that can deliver a customized measurement of natural elements for any given static location.

The mapping system assigns each area within a city a score between zero and one hundred, where higher scores indicate greater levels of biodiversity and access to green infrastructure. A lower NatureScore suggests that an area may have fewer trees, less natural vegetation, and fewer wetland areas, the resources whose absence negatively impacts the environment and the physical health of people living there. These lower NatureScore areas are often associated with higher air and water pollution levels, less shade and greenery, and a greater likelihood of flooding during heavy rains. This can mean higher stress levels, reduced quality of life, and shorter life spans for residents.

However, a lower NatureScore does not necessarily mean that an area is completely devoid of greenery or biodiversity and does not imply that the site is unsuitable for living. Rather, it highlights areas that can benefit from a targeted investment in green infrastructure and ecosystem restoration to promote a healthier, more sustainable urban environment.

With COVID-19 descending just as the co-founders got their company off the ground, it became their first test for studying the impacts of nearby nature on public health. So Jared, Bailey, and Minson abandoned their other early projects to focus on investigating the link between NatureScore and COVID-19. After analyzing data from the five hundred most populous American counties, they found an inverse relationship between NatureScore and the virus's spread—the higher the score of a particular area, the less COVID-19 appeared to affect nearby residents. While population density was the strongest predictor of cases per capita, NatureScore—even when controlled for population density—had a decisive, statistically significant impact.

Although we remain only at the cusp of understanding the range of applications of the rich connections between nature and longevity, the number of fields making sense of them quickly expands. The information from NatureScore models is just one example of how data-driven observations about nature are already improving analysis in healthcare, real estate, and financial research. Given the strong associations that NatureScore has shown with outcomes such as longevity and real estate prices, financial analysts could use it to improve their forecasting models. Total pension plans or annuity liability projections could be tweaked based on understanding the amount of nature near members' homes. Real estate portfolios or mortgage-backed securities could also be analyzed based on the current or anticipated proximity to nature.

In 2019, the multinational investment bank and finance corporation Citi launched a real estate development tool that would demonstrate the power of NatureScore to target new projects for the economically disadvantaged. City Builder is a free, data-driven platform designed to help investors, developers, and municipalities research Opportunity Zones, lower-income areas where the federal government offers tax benefits for real estate investments. The benefits aim to spur economic growth and job creation in communities that need it most. City Builder aggregates key social information such as population, demographics, education, land usage, and mobility scores to help developers and their financing partners make impactful investments. But the first iterations of the software failed to account for the all-important implications of nature—or, more often, its absence—in these neighborhoods. A recent report from the U.S. Environmental Protection Agency laid bare this oversight, finding that the greatest harms from climate change—including heat waves, poor air quality, and flooding—fall "disproportionately upon underserved communities [that] are least able to prepare for and recover from" them. So in 2021, City Builder added NatureScore to its other metrics. Now, NatureScore can help prioritize the delivery of green infrastructure and optimize investments in public health.

This application of the technology propelled Jared, fresh off the development of NatureScore, to take another step: making actionable the results of the company's analysis. The NatureScore Priority Index (NPI) leverages NatureScore to give researchers, nonprofits, municipalities, investors, and others a means for quickly identifying nature-deprived and socioeconomically deprived communities. Paired with research into nature-linked longevity, NPI allows advocates to model the health benefits that will accrue from specific investments in greenery, targeting the projects that will bring cities the most bang for their buck.

If city planners could properly value the health benefits of nature, they would have a much easier time persuading residents to invest in improving outdoor spaces. In the past, the focus of discussions about supporting, preserving, and developing urban nature has been unreasonably confined to aesthetics—how many trees are in one area compared to another, and how we can add more greenery—suggesting that the presence of nature in our neighborhoods is merely a luxury. An empirical, economics-based approach reveals that attitude to be a complacent fallacy. In its place, it substitutes indisputable measurements that allow residents to justify preserving urban nature even to the most skeptical taxpayers. Suppose products like NatureScore can demonstrate that a $500,000 investment in planting and maintaining trees or greening a street leads to increased property values that bring a city hundreds of thousands of dollars per year. In that case, it offers nature and all the residents benefiting from it a voice they previously lacked in our financially driven systems for allocating public resources.

The idea echoes that of renowned Cambridge University economist Professor Sir Partha Dasgupta, who the British government commissioned to write a comprehensive report on the financial worth of nature. Published in 2021, Dasgupta's findings present a compelling argument for why we must fundamentally shift how we measure and value the environment, calling for a redefinition of economic progress that considers the value of natural capital. Drawing on his decades of expertise in development economics, he challenges the prevailing models which long-deemed nature as a mere "externality." Instead, he proposes that we recognize its intrinsic and instrumental value. One of the report's key economic lessons is that investing in biodiversity and natural resources is morally right and makes long-term economic sense. This reality becomes undeniable when governments properly recognize the monetary value of ecosystem services and incorporate natural capital into national accounting systems.

Dasgupta's work is tested in cities, where space and real estate come at such a high cost that officials sometimes struggle to compel developers to include natural areas in their projects. Lacking the economic models necessary to incentivize such areas, cities face unique challenges in pursuing urban nature. But a comprehensive financial model that accounts for the value of nature in mitigating natural disasters can fundamentally change how we advocate for conservation efforts. Instead of an aesthetic or morally driven argument, preserving nature can suddenly be considered the only prudent decision in high-risk urban areas. By viewing the tangible benefits of urban nature in reducing flood and wildfire risks, alleviating heat waves, and contributing to improved air quality, sustainability measures can finally triumph.

Urban communities have quickly caught on to this idea. Barely a year old, the NPI is already being used to prioritize the delivery of green infrastructure and help inform public health policy in some of the U.S.'s most disadvantaged areas. One such community is Philadelphia, which boasts over 40,000 vacant lots. As the NPI reveals, these lots have some of the lowest NatureScores found anywhere in the city. If they could be greened, Philadelphia's mayor realized, they could transform these blemishes on the neighborhood into vibrant local resources.

The initiative that resulted, LandCare, has a mandate to "clean and green" the abandoned lots. It has become a model for transforming vacant lots into community assets, with more than 13,000 lots covering more than 16 million square feet of land in low- and moderate-income neighborhoods. It prioritizes those near schools and certain other developments, investing just $1,500 per lot to remove trash and debris, grade the land, plant new grass and a small number of trees, and perform regular maintenance. Some neighbors have taken it further, adding picnic tables and planting gardens in their new communal spaces. Where teenagers might have once tossed empty cans into once-neglected lots, they now feel ownership of the space.

In a study of the project by five physicians at the University of Pennsylvania, residents who saw their vacant lots greener experienced a 40 percent decrease in feelings of depression and worthlessness. Controlling for the different variables, the researchers found that cleaning up the trash alone didn't reduce negative emotions. Nature exposure, it turned out, was the key catalyst.

While it will take years for the full benefits of the green lots to be visible in longevity data, two effects quickly became clear: increased time spent outdoors and reductions in gun violence. Researchers from the University of Pennsylvania, UCLA, Rutgers, and the U.S. Forest Service, who compared crime data near the green lots eighteen months before and after the restorations were completed, came to these conclusions. They found that more than three-quarters of the residents significantly increased their use of outdoor spaces due to reduced perceptions of crime. Perhaps most compelling, gun violence in the surrounding neighborhoods fell by 29 percent. If other American cities with extensive tracts of vacant lots adopted the measures, this reduction would save hundreds of lives each year.

For all kinds of techno-naturalists, the results are a stunning testament to the power of modern ingenuity combined with mankind's most ancient necessity. A tree, some grass, regular maintenance, and a whole bunch of big data to target your efforts can solve some of our gravest and most intractable problems, increasing our life spans and the number of years to live a happy, healthy life.

Louisville, Kentucky's largest city, is taking Philadelphia's program further. After two decades of tree loss that saw overall canopy cover drop by 13 percent, the city is now conducting a $15 million, five-year clinical trial. It's the first of its kind, where urban greening is tested with all the rigor that the FDA would apply to a new pharmaceutical. But this time, the new drug is nature. First, the baseline health of 835 participants is thoroughly examined. Then, on nearly half of the participants' properties, the city will plant 8,000 trees and other shrubs, one of the largest projects of its kind in U.S. history. The other half, the control group, receives no new greenery. Researchers observe how those shrubs and trees impact a resident's health two years post-planting.

The study comes not a moment too soon. Wealthier neighborhoods in east Louisville have up to twice as many trees as poorer neighborhoods on the city's west side, where life expectancy lags its eastern counterpart by thirteen years.

More cities target their own innovative nature interventions every year, seeking to capture its benefits on human longevity. In 2009, a day in the life of a citizen of Albert Lea, Minnesota, looked like it would for many others living in small-town America. But this changed when Dan Buettner rolled into town. Dan's groundbreaking book *The Blue Zones* uncovered how some of the planet's most vibrant centenarians had remained so healthy over many years—through simple tweaks in their environments that encourage healthier habits. He'd discovered that trying to change one's behavior never works, but changing one's environment does, often *because* it unconsciously changes one's behavior. He sought to use this knowledge not as a prescription for others but to explore how everyday settings affect our health habits and overall well-being. By introducing simple adjustments that support healthy living, his dream was to bring these same life-prolonging benefits to middle America—one city at a time.

I met Dan over a video call while he was in Mexico. After a hectic tour promoting his latest book, which aims to put the insights he uncovered in *The Blue Zones* into practice, Dan ventured to a small enclave in Mexico where he could practice what he preached. There, surrounded by the beauty of nature and living simply as slower-paced locals do, Dan answers my questions in true Blue Zone style.

Albert Lea, with a population of about 18,000 people, became the first American city to sign on to the Blue Zones Vitality Project, seeking something of a longevity "makeover" by transforming roadsides into green, walkable pathways, providing more healthy menu options at local eateries; and encouraging nutritious snacks within school premises. Most notably, Albert Lea expanded its community gardens by 150 percent, with five new gardens spread throughout the town. Residents now access fresh vegetables and plants from the great outdoors—and foster meaningful connections over healthy produce. Within eighteen months, residents' body mass index (BMI) and healthcare expenses decreased and life expectancy for the average participant rose by 3.2 years.

I'm amazed when Dan tells me this. "How could it possibly happen so fast?" I ask.

"You get a representative sample of the city to take a life expectancy test at the beginning of the project," he says. He talks quickly, raising his eyebrows up and down his tan face like he's sharing something he's thinking about for the first time, even as the words flow effortlessly. "Then you ask them to take the test at the end," he says matter of factly, not quite answering my question. He excuses himself for a moment, turning to say a few words in Spanish to someone offscreen.

"There were other interventions," he adds, thinking through my question, "but I chalk up the community garden expansion as the most important one we did."

Dan's research into the Blue Zones has identified nine key factors that have increased longevity and well-being. Each is deceptively simple given their astonishing effects—tenets to live by like relieving stress; moving naturally throughout the day by commuting on foot, gardening, or doing housework; eating a plant-based diet; and fostering close connections with friends and family. While nature is not one of the so-called Power Nine, its presence appears to be a multiplying factor across the nine commandments. The five Blue Zones where people live longest around the world are Okinawa, an isolated island paradise in Japan; Loma Linda, a small California town; Ikaria, a Greek island in the Aegean Sea; Nicoya Peninsula, the largest peninsula in Costa Rica, and Barbagia, a region of Sardinia, the second-largest island in the Mediterranean Sea. When Dan reflects on them, he admits they have a tenth common denominator: "being outside." That's something, he says, that all Blue Zones have in common.

I ask Dan what it all means for the large cities, housing an increasing share of the world's growing population. "Should we turn around," I ask, "and run for the hills?"

He shakes his head. "It's a double-edged sword," he explains. "There's better access to healthcare and high-paying jobs in cities than in rural areas. And places like Singapore and Hong Kong produce the world's longest, healthiest life expectancies. But then there are places like Bangkok, where life expectancy is significantly lower." He pauses, tapping a finger on the desk, which his mic picks up. It reminds me of a woodpecker I'd hear outside my window in the summer mornings in Amsterdam. Dan says the glaring green-space discrepancies between Singapore, Hong Kong, and Bangkok could be a factor.

Almost half of Singapore and Hong Kong's lands are blanketed with lush foliage, while only 3 percent of Bangkok is reported to have greenery.

He turns to look offscreen again. *"¡Hola! ¿Cómo estás, Vecino?"* he shouts back to his neighbor, apparently through an open window. He tells me that the man is tending to his garden. "Diet, exercise, and stress play major roles in wellness," he says. "And as I think you appreciate well, nature's gifts play a crucial role in all of—"

A muffled voice cuts Dan off.

"*¡Ya voy! ¡Ya voy!*" he shouts, saying he's on his way. "I have to go help my neighbor in the garden." He smiles and waves. "Thank you for this!" And with a click, he's gone.

If the king of longevity himself agrees that the presence of nature could be somewhat of a litmus test for the health and life span of communities, then NatureScore could offer the alarm that its creators believed it to be, a way to identify where the action is most urgently needed. Of course, technology can only be the spark. As the enterprising people who brought picnic tables to the greened lots in Philadelphia understood, passionate users who use the technology will need to provide the fuel.

Ben Wilinsky, a champion of tree-planting projects around the world, has learned that public support is essential for the success of any urban greening program—but perhaps especially those in nature-deprived areas. Ben is no stranger to such areas, having lived in one of them— Englewood, Chicago—for three years, observing how its residents related to nature and the essential nature services they missed out on.

Ben came to Englewood, one of Chicago's poorest neighborhoods, after college, when he was tasked with teaching at the local middle school as a member of Teach for America (TFA). This nonprofit organization enlists college graduates of all academic majors to teach for two years in low-income public schools. The neighborhood was home to roughly 24,000 people, more than 90 percent African American, and its residents suffered from a disturbingly low average life expectancy—just sixty years. The cycle of disinvestment had built over the decades before Ben moved there, as the neighborhood gained infamy as Chicago's murder capital, with more yearly homicides committed than in any other neighborhood across the city. Shootings occurred on an almost daily basis.

Having never lived in a neighborhood as disadvantaged as Englewood, Ben was surprised to see how little it had in the way of parks and trees. "Whenever I drove across the city," he says, "from the rich north side of Chicago to Englewood on the south side, I would be confronted by how it got progressively less green. There was asphalt for as far as my eyes could see. I even coached football on the concrete." But the continuous reverberation of gunshots stayed with Ben the longest. "In a city with no trees, gunshots sound like someone dropping heavy two-by-fours into an empty metal dumpster," he says, wincing. "The sounds just keep echoing off the buildings and streets."

Ben shares his experience with me as we sit on a park bench in Williamsburg, Brooklyn. He has a thick mop of brown hair, transitioning to a three-day beard on the lower half of his face. Once tucked into his beige khakis, his checkered shirt sneaks out the side. Within minutes, we're interrupted by a torrential downpour, the first of several that week in an unseasonably warm end of September in New York. We run inside the nearest café.

I asked Ben if there were any parks in Englewood where he could take his students. "There was one," he says, "but it wasn't close, and the school's principal forbade us from going there. Too dangerous." He shrugs off the wet lanyard hanging around his neck—the pass for the conference that brought him to the city—putting it in his pocket. "Sometimes I wish things would change enough to take my students there."

Given his clear affection for his students—and his passion for mentoring them—why did he stop teaching? "Ultimately," he says, "even though I built these incredible relationships with individual kids, I didn't feel like I was serving the community as well as I could."

He feels differently about the work he ended up moving to. After stints in fundraising for community-based nonprofits, Ben serves as the director of partnerships and innovation at the Arbor Day Foundation

(ADF), one of the world's largest tree-planting organizations. Since 1972, ADF has worked to inspire people to plant, nurture, and celebrate trees, which its leaders and donors see as a keystone for healthier, happier neighborhoods. Since joining ADF over six years ago, Ben has been honored to return to Englewood to complete several tree-planting projects throughout the neighborhood. "It's a start," he says, "at building a better neighborhood." But he's far from finished.

As ADF's innovation leader, Ben evaluates and pursues forward-looking opportunities to bring the organization's mission to market. Every day he finds himself inspired by how ambitious that mission has grown. Over the last fifty years, ADF planted 100 million trees, which may seem like no small feat. But in 2022, in honor of its fiftieth anniversary, ADF announced it would plant 500 million trees over the next five years. The most exciting part of the plan, Ben says, is that the organization will target those tree-planting efforts in neighborhoods and forests with the greatest need. Ben was tasked with the challenge of identifying those areas. Struggling to find a rigorous answer was how he met Jared.

"There is an infinite amount of research that shows people who live near trees and green space are mentally and physically healthier than people without them," says Ben, "but until now, you couldn't see a tree's impact on a block-by-block radius—at least not in a way that's visual and easy to understand." NatureQuant changed that—for Ben and the entire ADF team. The proposed partnership got immediate support from the foundation's president. "Now we know how to target our work and how to talk about its impact."

The messaging, it turns out—explaining how a tree will impact the lives of residents and what role the ADF will play in maintaining it—is far more important than I would have realized. I'd have expected that anyone with space for one would want a tree—for its cooling benefits, impact filtering the air around your home, and beauty. But research indicates that my assumptions are off-base.

Christine Carmichael, an urban forestry scholar at the University of

Vermont and founder of Fair Forests Consulting, studied why a recent major tree-planting campaign in Detroit failed. During the four-year effort, nearly one-quarter of eligible residents submitted a "no-tree request," turning down free street trees. The nonprofit sponsoring the initiative had no idea why.

As Christine discovered, most of the "no-tree requests" could have been better interpreted as requests for more information. Some residents were wary of new trees because they'd spent years caring for street trees and vacant lots without help from the city. Growing up in Lansing, Michigan—some ninety miles west of Detroit—Christine had marveled how her hometown had artfully nurtured her neighborhood's street trees, transforming them into lush oases. But she quickly learned this was not the case in Detroit. Some residents who submitted a no-tree request wanted more trees, but not the large, shady trees that the nonprofit planned to plant, instead preferring ornamental trees that were easy to maintain. (One resident cited liking "those little trees with white blooms on them.") Others were concerned that Detroit would neglect the trees as soon as they were in the ground, a fear that was by no means unfounded—in 2014, years of population declines and cuts to Detroit's maintenance program had resulted in 20,000 dead or hazardous trees across the city.

Christine's study revealed that residents resisted the tree-planting process due to a lack of involvement in the decision-making—from species selection to planting locations to maintenance protocols. Ben learned from those findings that to be successful, the ADF would need to partner with residents throughout the planting process and the long term. It needed to share its insights into the quantifiable benefits that particular varieties of trees would provide—on heating costs, crime, air quality, flooding, longevity, and other measures—and then help ensure that whatever trees it agreed to plant with neighbors' input would be faithfully maintained.

Learning about how Ben used Christine's data, I remember back to the campfire at Clear Lake, where Jared shared his vision of a world where discussing these insights is commonplace. "Someday," he'd said, studying the canopy above us, "neighborhoods will be scored on nature access, just like crime rates, school performance, and economic production score them." At the time, his prophecy struck me as grandiose, but Jared's ideas came into focus in light of the role of quantitative measures in overcoming residents' anxieties about trees. He'd predicted that tax incentives would be provided for individuals to support green infrastructure and to engage with nature; doctors would draw on a patient's NatureScore to provide superior treatment; employers would encourage their workers to take breaks in nearby greenery; insurance companies would offer discounts on premiums for those who spend at least five hours per week outside.

"The impact of nature will be so hard to ignore," Minson had added, tossing a twig into the flames, "that people will be paying to go on guided 'nature retreats' in their own neighborhoods."

The day after we go to Clear Lake, Jared takes me to meet Scott Altenhoff, formerly an urban forester for Eugene, Oregon. Since 2022, Scott has led Oregon's Urban and Community Forestry Program. In this unusual but important role, Scott is tasked with supporting cities and community groups in planting and maintaining trees in urban areas.

When many are awakening to the disastrous impact of urban deforestation, Scott sees his work as promoting awareness about the need for more nature and the importance of growing tree canopies, especially in poorer neighborhoods. The job requires suits, ties, and difficult conversations with anyone, from lobbyists to developers to CEOs and politicians. When they allow, Scott prefers to converse on branches on the tops of trees.

Like the mayor of Eugene last month, I'm the latest person Scott has convinced to climb a tree. We're standing at the foot of a 130-foot Douglas Fir. I can feel my breaths quickening as I trace the trunk line until it disappears behind the thick pine needles. This is my first foray back into tree climbing since falling out of my friend's tree house at age eight, which by some miracle only bruised my ego.

Scott straps me into my harness, nearly lifting me off the ground when he checks it's secure, surprising me with his strength. He's clad in a Patagonia sweater, cargo pants, hiking boots, and a North Face cap that he quickly swaps for his helmet. Scott asks me if I'm nervous, and I shake my head no, even though I'm a terrible liar. He tells me not to worry. "I'll make sure you're safe," he says. In 1997, he adds, he spent more nights sleeping in "tree boats"—canvas hammocks typically hoisted 160 feet up a tree—than in a bed. I nod, taking a deep breath. Hearing that, I feel more confident, though I also lovingly wonder if this guy lost his mind.

Before working with the city of Eugene, Scott spent over a decade as a commercial arborist and forest surveyor throughout the Pacific Northwest. He has taught arboriculture and tree climbing to many up-and-coming arborists, eventually sitting on various committees and boards supporting community forestry.

He clicks my helmet closed, asks me to shake my head back and forth for good measure, and ensures I have all my ropes. I go through the motions of checking his ropes—a piece of climbing etiquette I've been taught—though I point out that I'm not confident I've checked correctly. I clip my microphone onto his shirt lapel, stringing the wire through the back of his shirt, plugging it into the recorder, and fastening it to the inside of his belt bag.

Halfway up the tree, we discovered I didn't do a good job. As he fumbles in his back for a sample kit he'd brought to test the tree bark, he pulls the wire out, and there's no recorder attached. At some point in our ascent, it had tumbled into the thick understory below. I planned to order a new one, but unbeknownst to me, after we finish our climb and he had dropped me off, Scott would return to the tree and search the brush, eventually finding the recorder and shuttling it back to me. I realized then how he ended up in urban forestry. He's a people person, and it offered the perfect sweet spot between helping people and nature.

The needles crunch underfoot as I secure my gear and walk toward the tree. When I blink, Scott has already ascended about twenty feet, looking half-monkey, half–Michael Jackson doing a vertical moonwalk. Now he looks down at me: "Your turn, Nadina! You got this!"

I flash him a smile and a reluctant thumbs-up. I gulp, staring up at the height of the colossal fir, the density of its branches like spokes on a wheel. How am I going to contort my body to get through these?

Vowing to give it my best shot, I use my arms and core to pull my legs around the bottom branch, brushing my chin against the bark. Then, with all my modest strength, I pull myself on top of the branch. Now dangling, crouching on a branch just a few feet above the ground, I go through the same process with the branch above me, this time with more force. When I'm a dozen feet up the tree, I breathe for what feels like the first time in minutes.

"Good stuff!" Scott yells down. "You're a natural!" I know he says this to everyone, but somehow it still helps.

Eventually, I made it to his branch. After I catch my breath, I asked him how he got involved in arboriculture.

He knew from years of tree climbing that he felt healthier and happier when he was close to nature, he tells me. But it wasn't until he met Vivek Shandas—the urban heat scholar whose story I discuss in Chapter 3—that the link became real. "Working with Vivek," he says, as he stands and looks up to visualize his next move, "bringing him down to Eugene from Portland to do an urban heat island assessment . . ." He trails off, taking a couple of breaths. "That opened my eyes to the critical disparities," he continues, "between the presence of nature and the connection with human health and well-being."

Before I can follow up, he's off again, and I follow. Too quickly this time. I lose my balance and slam my body into the trunk, the rope connecting to my harness snapping tight.

"Just relax," Scott says. "It's not a race!" His voice sounds so distant that I don't dare look up and try to find him. I catch myself apologizing to the tree—for the blunt force trauma and the lichen I had accidentally brushed off. "Let's try that again," I mutter, my left side now covered in gooey sap and fragrant needles.

This time, I make it onto the branch safely and feel a rush of relief, even though I'm developing a bruise on my left hip, a souvenir that will stay with me long after we descend, clambering up the knobby bark of the Douglas fir is a strangely calming experience. The crisp air brushes my skin and whisks any anxieties away. I climb higher and higher toward a bright blue sky stretching forever. I am so focused on my mission and insulated by the hush of the tree's needles that I forget about the stress and noise of the city that lies below.

Climbing closer to the trunk now, I'm surprised at the width between the branches I can scale. Each move—carefully selecting where to place my hand or foot next so as not to cause any more harm to myself or the tree—becomes almost subconscious.

Reaching the top, I'm rewarded with an unforgettable feeling of accomplishment, as if everything in the world has aligned for this moment. My eyes pan across the landscape, densely packed houses jutting down jagged roads built around the city's hills, separated by patches of forest; a soccer field where children who don't look much older than toddlers cluster around the ball, zigzagging around the area like a swarm of bees.

"Get comfy," Scott suggests. "We'll be here a while." I sit back in my harness, drop my hands from the ascender, and let my knees slack. A small branch—one of the only ones above us—cracks, showering me with moss, a clumsy squirrel accidentally making his presence known.

I watch in disbelief as Scott pulls a thermos from his bag, pours two cups of coffee, each with a lid, and passes me one.

"Are you serious?" I ask.

"I figured you probably didn't have your morning cup of joe yet." This guy was something else. So I thank him, and we sit side by side, dangling 130 feet up in the air, swaying gently in the breeze, sipping our coffee.

Eugene looks different from up here, a vibrant, verdant tapestry of life bursting from the undulating ground. We sit silently for a few minutes as tender gusts of wind rock us back and forth.

"I want this to be something everyone, young and old, can experience in their own city," Scott says. "Even if it's not in a tree, just being around them."

In recent years, Scott has become—for lack of a better word—obsessed with nature's profound impact on physical health and mental well-being. He argues that exposure to nature is particularly useful in poor neighborhoods because they may have higher levels of environmental pollution, which has negative health impacts. But it's also because, from an access perspective, residents in those neighborhoods may have fewer opportunities to travel to pursue recreation and leisure activities, which can affect the quality of life.

"I've been trying to convince people of the research for years," Scott says. Early in his career, he thought that simply publishing papers on nature's impact on health would be persuasive enough to secure the funds he needed to support community forestry. But he grew frustrated when he felt that his findings failed to persuade cities and states to make the investments they needed to. "When I discovered NatureScore, it felt like the Holy Grail," he adds. "After speaking in the abstract for so long, I could finally communicate *why* we needed more nature in an easily accessible way."

Now leading a statewide urban forestry program, Scott sees Nature-Score as an important tool in "connecting the dots." Previous GIS assessments had been too coarse, offering scant insight at the neighborhood or census-tract level. Still, NatureScore can be more granular, allowing him to connect cause and effect at the level of a lot. "The big question people want to be answered is: if I green this parcel or plant ten trees on this street, what effect will that have on the people who live right next to it—now and in ten, twenty, even forty years' time?"

Breaking down the data inspired the park and forestry teams across the state to make their big goals manageable. For the past two decades, for example, the park employees knew that their canopy status was falling short of state goals, and talking to many of them, Scott discovered they felt paralyzed by how far short the progress was. So using Nature-

Score, he calculated exactly how many trees could be planted in each park to meet Oregon's goals. That was when it hit home for park managers who could get them where they needed to be. Typically, they only needed to plant a couple dozen trees for a large park.

"So did they do it?" I ask.

Scott nods. He leans back, resting both his palms behind him on the bark. "The thing with trees," he says, "is that the timelines are so vast. They take decades to grow and, as we know, can live for hundreds of years. It's easy to feel like anything we do in the present is insignificant because the investment takes so long to pay off."

But that's the opportunity, I say, interrupting. "A little bit of work today—a little watering tomorrow—and we can shape the future."

"Now you get it," he says. "Each tree that starts as a sprout that you could stamp out with a single step is a beacon of what life can bring—a home for honeybees or a majestic sight to behold for passersby. By planting and caring for trees, we connect this moment to the generations that will survive us. They deserve to breathe the fresh air that we do."

I take a few deep breaths of that pine-scented air.

He asks if I'm done with my coffee, and I hand him the cup. I guess we should go down, he says, though he doesn't seem to mean it, his eyes lost on the horizon. "We've got a lot of work to do."

I take one last look at the tree-dotted city beneath me. The blazing shades of autumn are electric, emerald and crimson against auburn leaves in the shadows that have fallen to the ground. I try to imagine the view as it would look, speeding back through time, the networks of branches shrinking until each of the thousands of trees in view are just saplings and then vanishing completely, underground seeds of infinite potential.

As I lifted my sap-covered hand to the rope and opened the ascender, I placed my other hand over my belly to feel my own little seed, which had yet to show. An hour before we'd started our ascent, I'd learned the biggest news of my life: I was pregnant.

CHAPTER 9

Raising Future Naturalists

We cannot protect something we do not love; we
cannot love what we do not know, and we cannot
know what we do not see.
—Richard Louv

I STEP BEYOND THE trees and into the dense brush. When I crouch underneath the drapery of a golden weeping willow, I find a collection of stones arranged in an unnaturally perfect circle, as if by another child who discovered this sanctuary before me. I spend a few minutes sitting against the willow's sturdy trunk. Eventually, I return to the winding path and make my way around it. The only sounds are those of the leaves as my feet shift and push them aside, but in this small patch of forest behind my suburban childhood home, I know that if I listen carefully, I'm close enough to hear my mother calling me for dinner. My eyes widen as I take in my surroundings. Everything feels new to me, from the gentle sunlight peeking through the trees to the coolness of the air.

The rustling of bushes catches my attention. My heartbeat races, unsure of what mystery will confront me. But as I peek through a clearing between two trees, I see only the gentle sway of some ferns disturbed by a breeze.

With a sigh of relief, I forge ahead on my journey, overcome by the sheer marvel of this world. The towering trees seem to possess a familiarity with me, their immense presence captivating my senses.

A shower breaks out and drops patter against the leaves overhead. I come across a fallen pine next, its bark and twisted roots covered in liverworts. I feel connected to this decaying log as I trace its rough surface. Even as the rain intensifies, my hair stays mostly dry, the enormous trees shielding the forest's inhabitants from the outside world.

Though I imagine my parents yelling for me to come home, I cannot resist the urge to wander deeper into the woods, eager to see what hidden creatures the rain might summon forth. I spot a single mushroom peeking out of the ground. I crouch to inspect its scaly cap, picturing the slugs that would make a feast of it.

The longer I stay in the woods, the more I change. It was as if the forest was speaking to me, teaching me how to appreciate the small marvels I too often trample over or breeze past, oblivious. I find a dry, grassy strip and lay back on it with my eyes fixed on the gray sky above.

I had always dreamed of becoming a mother, too, one day, but when I discovered I was pregnant, the responsibility felt dizzying. The friends with whom I shared the news assured me everything would be fine. After all, they said, each of our parents managed to figure it out. Yet underneath my worry was a fear that I wasn't sure earlier generations of parents had reckoned with: Could I provide a meaningful life for my child in this urban age of screens and digital isolation?

To some, raising a child without technology is almost considered neglect, depriving them of the tools needed to handle the realities of the twenty-first century. In the eyes of many pediatricians, however, raising a child with too much technology creates developmental risks, including attention deficit/hyperactivity disorder (ADHD). Left unrestrained, screens can become like a drug; every television, smartphone, laptop, or tablet transmits an irresistible pull of energy that transports them away from reality—until they return, only to find themselves in a world of distraction and disconnection. Environmental researchers and sociologists have long posited that children have an innate affinity for nature and the outdoors, a pull that I feel strongly about. Cautiously venturing into the unknown, exploring a wild and expansive world filled with beauty, wonder, and enchantment, I remember studying the woods to discover new plants, animals, and seemingly secret places. Pebbles strewn along the shore of a creek turned into puzzles. Fallen leaves became a canvas for colorful mosaics. From watching a single blade of grass seize sunbeams to tasting the woody sweetness of honeydew raining from a maple tree to admiring an endless canvas of white, billowy "animals" in the clouds, the mysteries of nature seemed to feed my soul. Although I didn't have the words for it at the time, with every pine needle–crunching step, I experienced a newfound sense of freedom and control, discovering the balance between playing on my own and with friends, reconciling small details into abstract thoughts, or running away from a creature that startled me in a moment of thrilling chaos. Each step opened my imagination.

Based on my experience, one might think I grew up on the edge of Yosemite or Kruger National Park, which couldn't be further from the truth. While I certainly had the fortune to visit some majestic reserves in my childhood, the experiences that launched my awe for natural landscapes happened within the little tract of forest between my subdivision and the next.

Nature in cities and towns can hold a special allure for young hearts—in the tiniest bird perched on a branch, the squirrel searching for her next meal. The importance of these fragmented spaces—allowing us to raise children who appreciate their place within the kingdom of living things—is utterly incommensurate with their modest size. Amid the hustle of modern life and the transience of digital objects, these miniature oases moor our children to the intricate system that gives us our breath. Yet their tiny dimensions—the very quality that makes them fit seamlessly into our habitat—make them all too easy to erase.

When I was forced to return to my hometown to live out the pandemic in the house where I grew up, I could put this risk out of my mind for a time. I cheered to see that a few of the patches of forest remained, and I'd sometimes boot up my iNaturalist app to learn more about a new kind of clover or bush. I'd take walks at dawn through the small marsh where the neighborhoods' stormwater drained, hearing only the birds and my footsteps, watching the mist swirl, breathing in the musky air. One morning, as the world began to reopen, I heard a sound more common before the pandemic: the distant hum of heavy machinery. I followed it, winding through the subdivision's deliberately illogical streets, until I could hear the rhythm of engines, sticks snapping like firecrackers under a bulldozer's tracks. It was an unwelcome reminder of how, in our unending quest to dominate our environment, we have all too often pitted our most powerful marvels of engineering against nature.

If we are to give our young ones the best chance to succeed—and to cultivate a future for the generations after them in which humans live in a habitat that supports us—we must give them the tools to see past the dangerous fallacies that have set technology and nature in mortal combat: first, that nature is a perilous force we must curtail; and second, that technology is nature's enemy. Instead, the path to raising smart, healthy, and well-rounded children may lie in encouraging them to use the boundless human ingenuity that unleashes their appreciation for nature, which supports our ability to thrive. Contrary to popular belief, these technologies can facilitate information processing, spark creativity, and provide a platform for learning about nature's indispensable role in our mental, physical, and spiritual lives.

A central challenge in making technology nurture our connection with nature is ensuring that the innovations don't themselves become our focus. The excessive and compulsive use of screen devices, such as smartphones, tablets, computers, and video games, can distract from the outdoors while negatively affecting a child's physical and mental health and social and emotional development. Children are more vulnerable to screen addiction than adults due to their developing brains and lack of impulse control. Screens offer instant gratification, triggering the release of dopamine in the brain, and they can provide a sense of excitement and reward that other activities struggle to replicate.

Thus, looking to consumer technologies to build our connections with the outdoors, we cannot overlook screen addiction's psychological and physical impact on young minds. Excessive screen time can negatively impact a child's cognitive development, attention span, and ability to focus on important tasks. It can also contribute to poor sleep habits, affect physical health, and increase the risk of obesity. Furthermore, screen addiction can lead to emotional and social difficulties, including decreased empathy, communication problems, and a lack of social skills. This can impact children's overall social and emotional development, impeding their ability to form meaningful relationships and communicate effectively. Screen addiction can also interfere with the development of the prefrontal cortex, which controls decision-making, impulse control, and emotional regulation.

But in a world where addiction to mindless iPhone games and superficial social media apps are ushering these challenges into young people's lives, digital tools that nudge us to explore our curiosity about nature offer an essential gateway. Think of them like a dating app introducing you to a potential partner. While the app may spark your interest, what you and your match do on the date determines your compatibility. Similarly, technology can serve as an entree to outdoor exploration. Still, it's up to us to make the most of it—to nurture a lifelong interest and connection to the natural world.

I had spent years trying to understand how nature and technology could help us map, monitor, manage, and cultivate nature in our communities for survival and sanity. As "eco-anxiety" increasingly challenges our mental well-being, no goal has seemed more crucial to me. Though not a clinical diagnosis, eco-anxiety refers to the stress, fear, and worry experienced by a growing number of people in response to environmental issues, namely climate change. A 2020 survey by the American Psychological Association found that over two-thirds of Americans reported feeling these emotions, with young people being the most affected. Similarly, the first-ever extensive survey on eco-anxiety among 10,000 young individuals (aged sixteen to twenty-five) revealed that nearly 60 percent of respondents from the ten countries surveyed were extremely concerned about climate change.

With its ability to shape public perception, the mainstream media deserves a share of the blame. While it may make for exciting news, the tendency to focus on sensationalist stories of disasters, polarized political debates, and tales of environmental catastrophes can lead to an unhelpful sense of doom and gloom, which can exacerbate eco-anxiety, especially when the news isn't contextualized with opportunities we can pursue to mitigate these problems. While it's crucial to raise awareness about the state of the planet, continually bombarding people with stories of disaster and impending doom can leave individuals feeling powerless and overwhelmed.

As the research of Dr. Robert Zarr, the creator of Park Rx, revealed, nature has a magical way of soothing the soul. It can also offer peace and hope for children and teenagers facing eco-anxiety. Ironically, it is often what fuels our eco-anxiety—our natural environment—that can also help to alleviate it, with the benefits of ecotherapy offering the calmness and relaxation that cure eco-anxiety. The natural world provides children and adults with a sanctuary that can transport us from the stresses of daily life to a serene, calming place where we can connect with the earth and all that it has to offer. Being surrounded by the beauty of nature can awaken a sense of wonder and curiosity, igniting the desire to protect and preserve the earth that will ultimately enable us to actually do so. For many of us—especially children and teenagers—addicted to phones, app-based technology inspires environmental stewardship, creating hope for future generations.

Now newly pregnant, soon to birth a tiny fragment of the next generation, this academic fascination at the nexus of children, nature, and technology became something more. It was my mission as a mother.

Searching for the right tools to enrich my future child's connection with nature, I traveled nearly 6,000 miles across the Atlantic Ocean to meet Richard Louv, one of the world's foremost advocates for fostering a greater connection between children and nature.

In the 1990s, Rich published *FatherLove*, which helped bring new attention to men as parents, leading to reforms in how future government censuses were conducted, in parental leave policies that employers provided fathers, and the benefits for fathers in nonprofit and government programs assisting families. Into the early millennium, Rich continued to interview families and wrote extensive books on modern parenting styles. After traveling across the U.S., studying the family lives of over 3,000 parents and kids, he wasn't surprised to learn that they reported their experiences radically changing. But the change the subjects most often pointed to surprised him: the relationship between children and nature. *Last Child in the Woods*, Rich's seminal work, laid out the consequences of how the relationship had been irrevocably altered, maintaining that the closed school buildings and domestic, sedentary environments that defined childhood deprived future generations of innate benefits to their physical and psychological health. With alarming frequency, these children presented with symptoms of what Rich called Nature Deficit Disorder (NDD).

The new term was supposed to be a throw-away line. "I didn't even want 'Nature Deficit Disorder' on the book's cover," Rich admits, "but the publisher thought it was good marketing." Though he was afraid it might attract the wrong kind of attention—namely, doctors irked by a journalist attempting to identify a new disease—he later realized that it gave the book's thesis the needed emphasis. In the final chapter of *Last Child*, Rich speculated about what it would take to spark a revolution to rekindle an appreciation for the natural world and its healing power in children and their parents. At that point, it was only a far-off dream, a way of getting parents across the country to talk. But in the years after *Last Child*, something clicked as the book gained a cult following and found itself on the *New York Times* bestseller list several years after it was first published. Nature adventure clubs were popping up everywhere, and educators were starting to see the value in nature schools. Even pediatricians like Dr. Robert Zarr were getting on board, writing nature prescriptions and dispensing maps of local parks and playgrounds to his patients. Over a decade later, the movement continues to grow. What was once a mere hypothesis has given root to thousands of academic studies showing the profound impact of nature on children's health and well-being—and the consequences of its deficiency. Rich has been heralded as the pied piper, leading a generation to reconnect with the great outdoors, receiving awards such as the Audubon Medal, once bestowed upon E. O. Wilson and Rachel Carson.

Now seventy-three, reflecting on a career that remains as vibrant as ever—he's currently working on his eleventh book—Rich is overly modest, quick to highlight the role of other researchers in bringing nature into the fore. After we finish our pizza, Rich and I are ready for a hike. He tells me he has the perfect spot scouted out, nestled in the nearby mountains of Julian, California, just east of San Diego. As we drive a short while to Volcan Mountain, the late-autumn sun starts to descend over the blanket of rolling hills, pine trees, and green meadows that unfold around us, dotted with wild poppies and violet primrose.

At the preserve, an otherworldly landscape welcomes me: jagged rock formations and billowing stems of a mixed conifer forest. For those searching for a nature dose an hour outside of San Diego, the Volcan Mountain Wilderness Preserve fits the bill. The conservation area has diverse flora and fauna that echo in a symphony of sounds and smells. Meandering through the atmosphere filled with birdsong and the sour scent of pine, Rich and I stride to our heart's content. We're an odd couple: me, deep in the throes of first-trimester nausea, excusing myself periodically to take a quick rest and try to hold down the pizza rising in my throat, and Rich, stumbling over loose stones in what he calls his "fake cowboy boots."

It's a challenging ascent to the summit, and we slip and slide our way up, our backs baking in the sun's afternoon rays. Eventually, just short of the top, we decide to start the journey down, fearing that we may lose the daylight. As I turn, I look out over the trail's edge where the light catches the lips of the distant, oak-strewn hillsides, golden grasses swaying in the wind across the intervening meadows.

"On a clear day," Rich says, between breaths, "from the top, you can see all the way . . . to Mexico."

Over the years, some critics of Rich's work have cast his theories about the risks of inadequate nature exposure as "anti-technology." But he doesn't see the influx of technology as the reason kids aren't exploring the outdoors as much as a symptom of changing social conditions. Even one generation ago, Rich explains, the world was a different place. Kids roamed their neighborhoods freely, playing with friends until they got hungry. Today, with the perception of danger lurking around every corner, parents are understandably concerned about their children's safety. Combined with parents' overscheduled lifestyles—consumed by work and family commitments that leave them unable to supervise children outside—this fear too often restricts little ones' freedom to run along grassy fields, roll down hillsides, or bond with neighbors' cats on the sidewalks. Technology is used to numb, pacify, and entertain children with pent-up mental and physical energy—covering up the symptoms of a developmental need that nature used to fulfill.

However, the forces depriving children of the physical exploration they crave transcend parents' busy lives and fears about leaving kids unsupervised—they are also rooted in urban design. "If you tell your kid to take a walk in San Diego," Rich offers, "she's not gonna get very far before she runs into a freeway and needs to turn around." He argues that similar constraints frustrate children's ability to play in cities nationwide. "Too often," he says, "we build neighborhoods to provide easy access for drivers—adding double lanes and rows of parking—but completely fail to account for the perspectives and needs of children, who need safe sidewalks, narrow streets with slow traffic and protective crosswalks, and vibrant green areas to play."

The conditions limiting children's nature exposure paint a disturbing picture. The National Trust, a charity and membership organization for heritage conservation in England, Wales, and Northern Ireland, surveyed more than a thousand parents of young children to test Rich's assumption that kids are less engaged with the outdoors. The study found that today's youngsters spend significantly less time outside, with an average of four hours per week—nearly half the amount their parents had. It suggested that the cause was not parents' appreciation for the importance of nature but perhaps something rooted in our neighborhoods. After all, 96 percent of the parents studied recognized the value of their children's relationship with nature and believed that outdoor play was vital to fostering healthy growth. And yet, as another poll demonstrated, three-quarters of U.K. children now spend less time outdoors than prison inmates.

Without having the opportunity to integrate nature into their daily lives, urban children may begin to see the wilderness as something dark or mysterious rather than inviting and miraculous. When noise, smog, and pollution become more commonplace than trees, playgrounds, and grassy meadows, the lack of nature access for our youngest citizens can have devastating consequences.

While Rich cautions that NDD is not an official medical diagnosis, the evidence suggests it presents similarly to ADHD decreasing attention spans. Unsurprisingly, exposure to nature has significantly reduced individuals' risk of developing ADHD. For example, a Danish study of over 800,000 children born between 1992 and 2007 found that the more time participants spent in green spaces, the less likely they were to be diagnosed with ADHD.

As of 2016, 6.1 million American children aged two to seventeen have been diagnosed with ADHD, and about 60 percent of those cases will carry on into adulthood. Yet Rich argues that even those alarming figures don't capture the full damage of our lack of nature since inattention, impulsivity, and hyperactivity are only a few of the psychological symptoms of NDD. One of the core symptoms—a lack of appreciation for nature and its benefits—can lead to a range of troubling effects. When children cannot enjoy activities like swimming, hiking, and camping—when they find themselves bored by birdsong, a sunset, and the beauty of a blooming field—they can experience restlessness and depression. A decrease in physical activity, which diligent exercisers typically find most rewarding in the outdoors, is another common symptom, leading to other problems such as decreased motivation and weight gain. Partly as a result, a fifth of U.S. children are now obese, making them more susceptible to liver disease, diabetes, high cholesterol, high blood pressure, and mental health challenges like low confidence and self-esteem. Growing evidence shows that myopia, or nearsightedness, can be affected by time spent outdoors. An Australian study on twelve-year-olds revealed a correlation between higher levels of outdoor activity and lower rates of myopia. Other groundbreaking research highlighted the benefits of outdoor play on myopia progression in schoolchildren aged seven to twelve.

Luckily, unlike ADHD, the cure for NDD is radically simple: exposure to nature, which can be sparked by easier access. In general, greater access to nature in children yields improved moods; resilience in the face of stress; and greater impulse control, concentration, academic achievement, and problem-solving abilities. A 2005 study conducted by the American Institutes for Research showed, for instance, that engaging sixth graders in three outdoor education programs improved their conflict-resolution skills. In addition, consistent exposure to nature and interactions with nonhuman species are important ingredients for fostering ecological knowledge, identity, and ethics.

As I slide down a muddy patch on Volcan Mountain, I realize Rich is no longer beside me. I jerk my head around, beginning to fear I'm lost, only to find him crouched on the trail's edge a few feet back, aiming his smartphone over a flower. Its tiny, delicate petals form two circles of upturned, star-like blooms nestled within a halo of yellow and white stamen.

"What is it?" I ask.

He tells me he's trying to find out.

The computer vision goes to work. Within seconds, we have a hit: the Canary Island St. John's Wort, known as *Hypericum canariense.*

"Hmmm," Rich says, his knees cracking as he stands, "I've heard of this one. It's invasive and can potentially replace native plant diversity. Something to keep our eye on."

A few minutes later, stopping to drink some water, I ask him how much technology is enough—to grow our appreciation of nature but not remove us from its wonders.

He shrugs. "It depends where a person is coming from, I think. The right tech gets us outside, enriching our experience. The wrong tech locks us into a screen." He stretches, leaning back to look at the sky. "Let me put it this way," he adds. "The path to the woods is not through the metaverse."

A minute later, he asked me if I agreed. I tell him that I think it's a delicate dance. The goal should always be to use technology to restore our equilibrium with nature.

As I would later discover, a new area of research has demonstrated how nature-inspired technology can venture so far into artifice that it fails to offer the connection we need—what Dr. Peter Kahn, a psychology professor at the University of Washington, calls "technological nature." Unlike IoN innovations, which strengthen our relationship with our habitat, technological nature seeks to replace our habitat entirely, digitally representing the wild through VR simulations or plasma "windows" streaming outdoor scenery. While Peter has found that these digital recreations of nature can bring marginal benefits to people who also experience the real outdoors, they fail utterly as substitutes, offering almost none of the emotional, physical, and developmental benefits that nature brings. His insights confirm that as sophisticated as our technology may become, we must never forget that it can only help us thrive when we keep our habitat at the center of the frame.

Despite growing up outside themselves and wanting the same for their children, many parents trying to motivate their kids to experience the thrills of nature don't know where to start. Any parent who has urged their kids to use their bodies to play a sport or other physical game understands that technology can seem to be the enemy. For a generation raised with smartphones in their palms and videos accessible at the swipe of a finger, a simple game of hopscotch might quickly lose its novelty. This attitude too often pushed consumer tech gadgets and products designed to embrace the outdoors in opposite directions, like weapons in a culture war. Yet some nature lovers, swimming against the current, have embraced discrete technologies as groundbreaking tools to reengage people with our habitat.

Danish game designer Stine Kondrup is one of those innovators. Stine argues that we should not stop at using smartphones to educate ourselves about the ecological world. Rather, our technology can go even further, helping us solve nature's greatest secrets. Such is the vision for World Safari, the mobile augmented reality (MAR) game that she designed.

Like Pokémon GO, a MAR game that took the world by storm in 2016, World Safari uses immersive and location-specific content technology on your phone to encourage you to find creatures in the wild. Instead of roaming around your neighborhood for fantastic monsters, as in Pokémon GO, World Safari immerses you in the natural environment.

After Pokémon GO surged into our lives, gaining 230 million users in its first year, Leejiah Dorward, an ecologist at Oxford University, envisioned an app like World Safari. She and her colleagues from Cambridge, the United Nations Environmental Program World Conservation Monitoring Centre, and University College London recognized the power of Pokémon GO to draw people out of their homes, and they wondered if the technology could be harnessed positively and proactively to reconnect the public to nature.

In just six days, Dorward found, Pokémon GO players had collected as much data on virtual animals as naturalists have collected about our planet's real animals in four hundred years. She and her colleagues reasoned that following the model of Pokémon GO, games that encouraged users to look for real species could provide a powerful tool for education and engagement.

Little did they know, it was already happening. While Pokémon GO grabbed the headlines, World Safari quietly entered the market.

The rise of Pokémon GO allowed Stine and her team to listen to feedback from that game's users and iterate new and improved features on their own. "There were a lot of stories about accidents and injuries sustained when Pokémon GO users tried capturing the most sought-after species," Stine tells me. "They were clearly engaged in the gameplay but also dangerously distracted." She sips a cappuccino, brushing her strawberry blond hair out of her eyes, letting it fall down her back, where it reaches nearly to her waist.

In light of the accidents that Pokémon GO players suffered, Stine and her team focused on creating hyperlocal content that required you to get instructions via your phone and then put it away, completing the assignment without it. By roving different neighborhoods, players can unlock new tasks that involve exploring the ecology around them. Instead of colliding into trees, World Safari users must immerse themselves in their environment.

"It allows people who never realized they were already in nature," Stine explains, "to consider the millions of relationships in the soil beneath their feet that make their lives possible." Stine reminds me of the implicit bias many of us harbor against urban nature, the perception that it isn't "real" nature. When we think of nature, we tend to picture vast forests, towering mountains, and unspoiled wilderness. In doing so, we forget that the wild exists all around us, even in our bustling cities.

The trees that line our streets, the parks where children play, and the birds that flutter above us are all part of the urban ecosystem, shaped by human influence but rooted in the same ecological systems and processes as any other environment.

The early research suggests MAR games have huge potential. A study from one of the top pediatric hospitals in America found MAR games help make children more social, make their routines more meaningful, help develop positive emotions, and motivate them to explore their surroundings. Several other studies have revealed that MAR games benefit physical health and cognitive performance since they increase players' daily activity levels and time spent outdoors.

Those benefits are magnified when these games direct our attention to nature, as World Safari does. In a landmark 2018 study, Amy Kamarainen, an ecosystem scientist at Harvard's Graduate School of Education, examined the opportunities of MAR games in environmental education. Kamarainen recognized promising applications in helping novice naturalists observe objects, patterns, or phenomena they might not otherwise notice—for example, the layer of silty residue left in the wake of spring floods, the intricately carved nesting cavities crafted by woodpeckers, or the paths of trampled brush cut by deer. Kamarainen and her colleagues found that because scientific observation is much more challenging than many think, MAR games that emphasize features of users' environments provide the opportunity to train our ecological eye.

As World Safari drives home, the natural world is a playground full of wonders waiting to be discovered. It's where children can unleash their imagination, channel their creativity, and explore the unknown. As another phone-based technology that encourages "nature hunts" would illustrate, such games also give children access to an entirely new realm of interactive engagement with the environment around them, one that unlocks breakthroughs that can heal our environment and alleviate eco-anxiety. Beyond simply fostering curiosity toward nature and instilling a sense of wonder that may last a lifetime, such tools empower children to take action as stewards of the environment.

When I hear a twig crack, I try not to gasp. Instead, I hold my breath. The leaves begin to rustle. He's coming.

I'm ten years old, sitting under a massive oak tree behind my house, waiting for squirrels. It didn't take long, maybe only fifteen minutes, before I heard one scurry down to the forest floor.

It was a game I would play: collecting acorns, aligning them like an all-you-can-eat buffet, and waiting. I'd stay frozen for as long as it took for my hungry customers to show up.

The eastern gray squirrel eyed the neatly patterned food with suspicion, his big black beady eyes darting around the brush. He sprinted forward, his fluffy tail—as wide as his body—dancing behind him. He grabs the first acorn with tiny T-rex paws and carries it to the base of the tree he came from. He pries off the acorn's cap in rapid succession, turns it over, and munches it up. My mom would kill me if I ate anything that fast.

I watch the squirrel devour two more, then grab six others and bury them. Eight acorns. A new high score.

Looking at the little holes the squirrel dug and seeing no pattern, I wonder how he'll ever find the acorns he stored. As an ecology student, I will later learn that squirrels are primary contributors in the regeneration and dispersal of oaks, as they fail to recover nearly three-quarters of the acorns they bury.

Squirrel-watching taught me the importance of patience. And although I'm sure that squirrels are much more adept at hunting for acorns than I was, preparing these buffets made me feel like I was giving back.

Two years later, I got my first phone. Soon, squirrel-watching would be replaced with playing Bejeweled—for hours and hours on end. Not only will I stop watching squirrels, but I begin to barely notice them. Even when walking to school along forest trails, I won't dare glance away from my screen, determined to get a high score of a different kind.

Nearly a decade later, I pay attention to squirrels again. In Toronto, where I'd study ecology in college, you'd need to watch your lunch around squirrels. The city creatures have learned to be opportunistic feeders, eating anything from french fries out of your hand to pasta salad scraps left in the trash. Their aggressive eating habits would pique my interest, and as I'd begin to study their movements across the park, I'd become fascinated by their color. These urban squirrels aren't the pale gray I'd remember from childhood but a dark shade of black.

As I learned, the eastern gray squirrel consists of two color morphs: gray and black. But how, I wondered, do different-colored squirrels come to dominate different areas?

As it turns out, the question has baffled scientists as well. In search of answers, I stumbled on the SquirrelMapper project. Using the iNaturalist app, squirrel-watchers like me can upload photos of our best sightings, classify the coat color, and play interactive games like Find the Squirrel. The data we submit helps researchers determine the conditions in which each morph fares best.

The scientists who launched the project have analyzed nearly 160,000 squirrel observations with locations and coat color classifications by involving the community. Even the most dedicated squirrel watchers couldn't clock the hours necessary to build such a dataset. What was once a mystery has begun to reveal itself: black morphs are likely more populous in cities because they're less camouflaged against the palette of grays that comprise the urban background. Unlike the gray morphs, which are overrepresented in roadkill, the black morph squirrels are more easily seen—and therefore protected—by humans.

As I would come to discover years after my renewed appreciation for observing squirrels, the same technology that reinspired my love of nature would kickstart an international movement, not only to observe squirrels but the millions of other species with whom we share our cities.

Community science staff, Lila Higgins at the Natural History Museum of Los Angeles County (NHM) and Alison Young and Rebecca Johnson at the California Academy of Sciences (CAS), are three researchers leading that charge.

In 2016, they founded the City Nature Challenge, an annual four-day global "bio blitz" nature hunt to record and protect wildlife in urban areas. In its first year, Alison and her colleagues launched a head-to-head competition between Los Angeles and San Francisco residents. Participants competed over the iNaturalist app to see how much biodiversity they could map. In the Challenge's four days, they logged over 20,000 observations.

The event went national the following year. In 2018, it went international. Then, it went viral. In only seven years, Alison and her colleagues grew a quirky idea into a global event in over 480 cities across forty-four countries. The participants of the 2023 edition of the Challenge observed nearly 1.9 million organisms, a nearly hundredfold increase from the Challenge's first edition. Millions of organisms were identified as members of more than 57,000 different species. The 66,000 children and their parents that participated outnumbered the workforces of the largest municipal park systems in the world.

Nearly 5 percent of the species documented were considered rare, threatened, or endangered, proof of life's persistence that may be a sign of hope for conservationists and the eco-anxious. When I ask Alison what inspires her to continue to invest in this event that has no doubt become as demanding as a second full-time job, she points to these sightings and the potential they promise for ecological restoration.

"During the 2020 Challenge," she recalls, "a fourteen-year-old girl identified a 'white-spotted slimy salamander' under a log in Arlington, Virginia." I set down my colorful beanie on the bench where I've rendezvoused in San Francisco's Golden Gate Park, home to the California Academy of Sciences—the birthplace of the Challenge. I breathe in the unseasonably cool air, blowing it out through pursed lips as if making smoke rings.

Even though the salamander wasn't endangered or threatened, she explains, it hadn't been recorded in the county since 1977, and most scientists assumed it had left the area.

Supported by the iNaturalist community, Alison published her findings in a top scientific journal, providing valuable data to researchers studying amphibian habitat changes. When the local news media later realized that a young girl had contributed a nugget of insight that had eluded ecologists for decades, they asked her if she'd always been interested in nature. The teen shrugged, saying it was her dad's idea to participate. He thought I spent too much time on my phone, she said. Ironically, with the help of Alison's competition, that same phone now engaged the girl in her natural surroundings, eventually leading to her admission into two college zoology programs.

Alison has hundreds of similar stories of young stewards discovering their love for the natural world surrounding them. She chuckles, telling me of last year's Challenge when a nine-year-old found a pill bug. Although native to the Bay Area, the bug hadn't been seen in so long that it was believed to be extinct. Now, city officials are taking steps to protect pill bug habitat so it won't be another eighty years before one is spotted again.

The observations of urban plant and animal life connecting a new generation with the wonders of the outdoors will benefit even kids who already deeply appreciate nature long after they have grown up. By collecting valuable data on our planet's biodiversity—with iNaturalist observations already underlying five hundred peer-reviewed articles—these observations inform local land management decisions, helping city managers design more biodiverse spaces to help plants and animals thrive for decades.

Community groups now lead some of this work, using the biodiversity heat maps provided by iNaturalist to inform and inspire their interventions. One example followed a sighting during the 2019 Challenge of the highly endangered *Erica verticillata* along stream banks in the southern suburbs of Cape Town. Local organizers banded together in the weeks afterward to save the hardy flower from the precipice of extinction. Two years later, *Erica verticillata* flooded those same stream banks.

But nothing could have prepared Alison for the 2022 winner that smacked Cape Town—then the City Nature Challenge champion for three years running—from its throne: La Paz, the capital of Bolivia. Nearly doubling Cape Town's observations, La Paz made a whopping 137,345 observations in that year's Challenge, comprising a stunning 5,320 species, shattering all existing records. La Paz had previously participated in 2019, finishing second, third, and eighth in the observations, participants, and species categories, respectively; the COVID-19 pandemic frustrated efforts in 2020 and 2021. Desperate to make a comeback, La Paz enlisted the youngest of its citizens—thousands of students from sixty-eight schools across the city's eight boroughs. Its leaders coordinated with a university and the city's natural history museum to have the Wildlife Conservation Society train the students. The young students embraced the initiative, acting as biodiversity foot soldiers who worked day and night to put their city's wildlife on the map. The strategy proved so successful that, in 2023, La Paz won again.

La Paz is a dynamic metropolitan area with a rugged landscape ranging from foothill forests at 400 meters above sea level to the iconic Illimani Andean peak, standing proudly at 6,450 meters. The city's range of habitats creates unique microclimates that support lush biodiversity. But deforestation and habitat fragmentation in the Cotapata National Park that borders La Paz had led some conservationists to conclude that it was too late to save some of the city's most treasured species, which hadn't been spotted in so long that many naturalists hardly knew what they looked like.

The massive effort proved skeptics wrong. Students turned up dozens of plants and animals that people thought had vanished from the region. One of the most remarkable discoveries was a *Phylodryas boliviana* snake. The elusive creature, native to Bolivia, had only been seen three times in over two decades of research. Equally impressive was the collection and sighting of a small wild blackberry, *Rubus conchyliatus*, which hadn't been observed in La Paz for over a century. The Challenge had not only generated massive enthusiasm for ecological exploration among children who had previously expressed indifference about nature, but it unlocked scientific secrets that drove them to engage with the outdoors in various ways.

The success in La Paz marked a turning point for Alison, who had become stretched thin by the demands of her research job at the California Academy of Sciences combined with the surging popularity of the Challenge. It showed her that the divided duties, which sometimes felt like they pulled her in opposite directions, were essential to each other's success. La Paz demonstrated that inspiring children to appreciate the magic that Alison encountered studying nature in her day job would be the only way to guarantee that research institutes persisted. And that the best way to introduce children to nature's mysteries was through a tool they were already comfortable with.

As I watch a Frisbee spin back and forth between teenagers in Golden Gate Park, Alison admits that she initially felt uncomfortable making iNaturalist a centerpiece of the Challenge. Still, her co-founders convinced her that the app would draw to the effort the allies the project needed in order to succeed. "It felt counterintuitive," she says, "to tell people to discover nature by taking out their phones. But when I saw how the Challenge inspired the students in La Paz, I knew that it was teaching them to see the creatures that many overlooked—or looked past—every day."

Like when you look through a microscope that's out of focus, I offer. One turn of the knob and the formless blob becomes an intricate cell.

In the dirt parking lot at the foot of Volcan Mountain, Rich and I gaze up at its hunched form, half patched with bare desert and low grasses, the other half rippled with trees of varying heights. As I try to envision the view from where I stood just an hour before, I feel the cold sweat from the journey trickling down my back. The sun has only just begun to set, leaving streaks of orange and pink trailing like fingers across the sky.

I imagine how the landscape might have transitioned backward and forward through time, forests overtaking the rocks and receding in years of drought, while the mountain's silhouette stood unchanged. If I squint and let my thoughts run, I can almost envision how the generations of plants might dance across the dominating ridge—a dense grove of conifers overtaking a sunny outcropping after decades of smaller bushes depositing enough soil to sustain them, another patch of trees that had grown too tall disintegrating gradually in the dry wind, only to fuel a later generation. Over tens of thousands of years, as these green blankets grow and recede, the ridge softens a hair, sagging as the tectonic plates below inch apart.

Strangely, this contrast between the flux of ecological time and the rigidity of its geological counterpart feels like a metaphor for the tension that first led me to approach Rich—between nature's deliberate movements and technology's frenzied chaos. I had my own reasons for trying to resolve this tension. I was intimidated by how to raise a child who could be guided by both forces at the center of our world, and I thought that Rich might help me find the answer. Now, if only for a moment, I can glimpse the forces in equipoise—the stillness and the movement perfectly balanced.

As if reading my mind, Rich breaks our silence to recall a story he'd been told by a seasoned instructor who had the task of training fledgling pilots for their first seafaring journeys on their own. The man, Rich says, spoke of two types of students he'd met in his years as an educator: one faction, skilled with machines and technology after countless hours playing and replaying video games, could easily adapt its knack for gaming to master a ship's electronics systems, predicting how to steer the optimal course with great precision; the other breed, outdoor buffs, fully conversant with nature's bewitching dynamics, intuitively knew where dangers such as shallow waters lurked. The groups bickered relentlessly whenever the ship grew lost in the fog on simulations miles away from the shore. Then one day, as the weeks-long training operation began to wind down, they could communicate *why* they believed what they did. From then on, no matter the weather, the crew would never lose its way.

The instructor's assertions were not unfounded. Across a range of contexts, social scientists have established the importance of being able to experience reality in both imaginative and logical ways, revealing that a multifaceted view often facilitates the clearest understanding of any given topic. Debate continues to surround the impact of technology on modern generations' sense of logic and imagination, with some hoping for an enlightened future where innovations spur superhuman intelligence freed from our physical constraints. Other researchers, such as Professor Mark Bauerlein, fear that our growing reliance on tools developed by someone else is producing the "dumbest generation" yet.

Rich offers a third possibility, the "hybrid mind." Achieving it requires raising the next generation to be tech-savvy and knowledgeable and have an uninhibited appreciation for the connections embedded in nature. He describes a delicate harmony that marries both the digital world and the outdoors, the ultimate convergence of our species' potential as an animal that evolved an unusual ability to build tools for itself. Children who master the hybrid mindset will understand their place in society, grow deeply in love with nature, and hone the problem-solving skills that help us through life's greatest challenges.

The theory, in other words, asks us to nurture tomorrow's leaders by giving them both the digital and natural tools to be successful in today's rapidly changing world—the one they'll be tasked with fixing.

We must heed this call for my son's future, your future, and every beam of life this planet brings forth.

"This hybrid mind," I ask Rich when he drops me off at the pizza restaurant where we'd started our journey that afternoon, "can altruism lead us there?"

Yes, it was a rhetorical question. But I wasn't prepared for Rich to laugh. As Rich seems to know well, the truth is that we humans are creatures of self-interest, biologically wired to protect our futures. As animals who learned to thrive using tools such as spears, fire, and eventually agriculture, we see the world through a lens of selfishness, forever alert to ways we can exploit the resources around us. As our population boomed and our tools grew ever stronger in recent centuries, our exploitation actions have nudged our entire species to the brink.

Now, peering over the cliff's edge, we are awakening to the immense value that nature holds within it—value that, day by day, we continue to destroy. By leveraging the power of nature, like parks and gardens, we can successfully manage stormwater runoff, reduce pollution levels, and create resilient cities that are prepared to cope with climate change, all while reducing stress and fostering the social connections we crave.

Someday I hope that our children will be able to argue that this whole book—that my career as an ecological engineer, in fact—entirely missed the point. Nature's value should not be reduced to what it does for us. That quantifying the benefit of a stream or a hillside—things altogether precious for their own sake—is as illogical as trying to bottle a rainbow.

Indeed, the sensations we feel when walking through an ancient forest, the spectacle of witnessing a shooting star, are beyond the capacity of explanation in so many ways. Future generations would say that we must marvel at the grandeur of our natural world without commodifying it or even trying to reduce it to the incapable terms we call language. In today's unnatural society, however, we must first show each other what our habitat means to us. For example, how two decades after Peter planted a community garden, the average health span of his neighbors has increased by ten years. How just a year after Marina removed the tiles from her backyard, her house suffered significantly less flood damage than her neighbor next door. How Shekar, who cleared the brush from his property, was the only one on the street whose house did not burn. How Sadie taught kids to play an augmented reality game for nature conservation, and now they come home with better report cards. How Moriko's company, which mandates daily forest walks, has employees ten times more productive.

To someday appreciate nature on its own terms, we must first bridge the gaping hole between ourselves and our environment by learning to name its manifold and highly tangible benefits. Imagine it this way: I hope to tell my son when he admonishes me for the self-centeredness of my perspective: a few years after you were born, I decided to go to the gym for a selfish reason—to work on my beach bod. But as I continued training, I discovered how tending to my physical well-being elevated those around me—it made me a more loving partner, patient mother, generous neighbor, and dedicated ecologist. In the years before you can remember, we'd grown so far from nature that we needed to follow a similar course. What started as a selfish attempt to quantify its benefits transported us to a place where we can now understand something so much more powerful—an intrinsic connection to our habitat that sparks transformations across communities far and wide.

The Internet of Nature (IoN) was never the destination. It was merely a stepping stone, a meandering path that led us to a place where we could appreciate the earth's beauty without the taint of economic motives—a place where a towering tree and a wriggling snake need no justification other than our love. The day our IoN becomes obsolete will be the day it finally succeeds.

The Nature of Our Cities

YOUR EYES FLUTTER OPEN, a bird's chirp welcoming you to the new day. It's the alarm clock you lacked for so many years, with the light pollution throwing off the birds' rhythms. But now it's back—and thriving.

You part the curtain and catch a glimpse of the three young saplings planted on your street last week, making a mental note to remind your son to water them after school. Since its humble beginnings on the west coast of Africa, the tree-growing campaign that earned citizen planters micropayments has been replicated around the globe. Your nine-year-old is its newest convert, unable to believe his luck in getting paid for what he once did purely for fun.

But because your little eco-hustler could sleep through a chorus of a thousand honking geese, you wake him. Soon, you're walking him to school, the first of many strolls you'll take on this unusually warm day. Even at 8:30 A.M., the sun sears your skin as you cross the street, navigating between the cones that rise to protect your path when the signal turns. But then, the mosaic of tree crowns on the trail that cuts through your neighborhood embraces you, and the air cools. You kneel by the saplings, feeling the surface of the soil.

"It's actually pretty moist," your son says, his fingers already inches deep. The algorithm in your sandals, fed by a citywide soil sensor network, confirms his observation, and you realize he's become a keen enough observer that you can delete it.

Since the municipality reimagined its approach to ecology, its trees are the least of its success. A decade ago, unable to ignore any longer the failures of its discrete approach to city services, the city restructured. Now it requires every department to team up and lift the hefty burden that a single agency—the Department of Parks and Recreation—had once tried to carry.

It isn't just the city oversight that changed. Disconnection was, it seemed, how we approached everything before the Flood. We no longer try to build our houses higher than our neighbors, hoping the water can be someone else's problem. Now, we know how we can all stay dry.

When you tried to explain the old days to your son yesterday, he didn't believe you. "If you couldn't predict the season's rainfall," he said, "then how are you supposed to shop for clothes?" He gaped like you were some kind of monster.

Even though you understand that the time will come when he learns about history, you are grateful every day that he doesn't have to know what it was for a city to be wiped off the map.

He brushes his soil-encrusted hands on his pants. "Let's check again tomorrow."

You nod, trying not to focus on the smear of dirt—no doubt the first of many he'll come home with this week from his nature-based school. Three-quarters of their learning happens outside, and today's curriculum is taking part in a regional "bio blitz," where students find, document, and identify as many species as possible through a computer vision–powered app.

You round the corner and see the school. Your son's teacher is waiting outside with her students, so you remind him to be kind to others and tell him goodbye for the day.

The school has the same name as yours, but everything else has changed. Here, thick shrubs, wild grasses, flowing water, mulched paths, and a wall of trees not only invite students to use their imagination but keep them cool and breathe purified air. The structure that was your school still stands on the other end of town, an eroding concrete husk in the Millennium District. These days, that whole neighborhood lies empty except for the Museum of Modern Design, where older children gawk at the follies of their grandparents and learn about the year that everything changed.

Perhaps someday soon your son will be ready for that. You've planned to take him camping—not in Yellowstone, but in a park outside the subway station, just five stops from your house. Foxes had been spotted last week, and a colleague mentioned that she saw at least a hundred constellations across the night sky. You remembered constellations from childhood, but only two were visible from the city for many years. As everyone in your ether-sphere has been streaming their snaps the past few days, you've felt like you were missing out.

"Can you take us?" An exuberant voice interrupts your thoughts, reaching out with her camera.

You nod at the woman and look around to find the other half of "us." For a moment, you're confused. She points to the sprawling pin oak behind her, and then you recognize the tree: Oakly, the talking oak. She rose to fame a decade ago after we learned the hard way that something had to change. Plugged into tree and soil sensors, she was programmed to continuously translate her data into perceptual messages, captivating an entire generation who followed her stream. You were one of her fans, and you learned that if you can relate to something that speaks your language, you're more likely to show empathy for it and perhaps even care for it more. Years later, Oakly confirmed this theory. She has more followers than all the 2034 Olympic teams for hover sports *combined*.

You happily snap the data and leave the woman to stream it. You walk off toward the park, circling the marsh that doubles as a reservoir during particularly wet seasons, swelling to a 40-foot-deep pond in which the cultivation of marine plants clears road salt before the water is treated. It's only a few feet deep right now, and the thunderstorms coming all week will be welcome.

The park is one of many. Initially, the city planned to build one within 300 meters of every home, starting with the nature-deprived neighborhoods. But as people began to use them around the clock, in shared backyards of sorts, for meeting their neighbors, the idea of having only one within 300 meters became laughable.

You pass a group of young women setting a table for al fresco dining later this evening. An elderly man tends his plot in the community garden, taste-testing most of his harvest. New moms and their newborns settle into a circle of boulders in the prairie grass for their weekly support group. Two young teens lie head-to-head, sharing their thought-link. You hear the distant laughter of children, no doubt excited about their bio blitz finds.

"A pedometer," a girl shrieks. She peels a red piece of dirty plastic from a hole she dug in the soil.

"A fossil?" her friend asks.

She shakes her head no.

You chuckle, impressed that the girl knew what she'd stumbled upon. She must have old parents.

The truth is, not unlike your parents with their pedometers; you used to track your nature exposure on a smartphone app, which nudged you if you didn't meet your daily dose. A few years ago, you realized it was obsolete. The patterns of everyday life had changed so drastically that not a week went by when you didn't triple the app's target.

Though you were glad to have the device, you were even happier to get rid of it. Like so much of the technology that used to be installed in every home—to measure groundwater levels, pump out gusts of cold air, and automatically swing the blinds closed when the sun outran the cooling system—it was a bridge that took you where you needed to be. And no matter what, you weren't going back.

ACKNOWLEDGMENTS

When I set out to write this book, I never thought I'd share this experience with a baby in my belly. Yet when I discovered I was pregnant a few days after signing my book contract, the project and my son, Luca, became inextricably linked. Whether I was driving through Sonoma Valley, telling Luca about the interview I was about to embark on; learning about the latest technological marvels on the streets of Amsterdam; or pulling all-nighters behind my desk to finish chapters, sharing this experience with him has added a profound depth to it all. Because we went through the whole process together, it felt fitting that Luca came into the world on the day I handed in my manuscript, albeit much sooner than we expected. Even though I didn't realize it when I embarked on the project, I couldn't have written this book without you, Luca. This book is dedicated to you.

My agents, Gail Ross and Ethan Bassoff of William Morris Endeavor, fell in love with this book when others doubted it. You encouraged me to think more, write more, imagine more—and push past the boundaries of the IoN and then push some more. For that, I am eternally grateful.

My publishing editor, Matt Harper of Mariner Books at Harper-Collins, gave this book its perfect home, and I'm so happy that he did. You trusted me when we disagreed and supported me when things got tough. You helped me dare to call myself a writer and make my lifelong dream of becoming a published author come true.

My editor and friend David Lamb believed in this book when it was a collection of scribbles in a worn-out notebook. You guided me through this process from start to finish. You believed in it when I didn't—and I don't know how I would have completed it without you. Through hours on the phone, we made this book everything it could be. Thank you.

There are many mentors and colleagues without whom I'd have never dared to write this book: Francesco Pilla, Carlo Ratti, Fábio Duarte, Marcus Collier, Cecil Konijnendijk, Ian Hanou, John Judge, Thomas Randrup, Jared Hanley, Chris Minson, Chris Bailey, Eva Gladek, Thomas Crowther, Clara Rowe, Richard Louv, Gil Penalosa, David Miller, Adrian Benepe, Dirk van Riel, Matthew Wells, Dan Lambe, Ben Wilinsky, Timothy Beatley, Max Lerner, Lucy Almond, Menno Schilthuizen, Marcel Steegh, Willy Detiger, Andrew Hirons, Joost Verhagen, Josh Behounek, Vivek Shandas, Jennifer Walsh, Monica Olsen, Ash Welch, Eric Ralls, Tim Rademacher, Yvonne Lynch, and the many other brilliant thinkers who have taught me so much. I hope those of you I'm forgetting will forgive me. Know that I haven't forgotten the way that you nurtured my ideas.

I've had the honor to meet, interview, and befriend hundreds of innovators within the IoN who generously shared their stories and wisdom with me. Your insights and experiences, some of which you'd never shared with anyone, were essential to creating this book. I'm deeply moved to have had the privilege to tell your story.

Every ecologist today who, like me, has the privilege of writing about the natural world stands upon the shoulders of giants: gifted environmental scientists, technology wizards, urban planning gurus, and nature writers. I am humbled to build upon your work and contribute to a growing body of knowledge. I invite all readers to join our mission and help foster vibrant ecosystems for everyone. We need all hands on deck.

In no particular order, my closest friends and confidantes—Riannon, Caitie, Nikki, Chrissy, Alice, Nadia, Ally, Esmee, and Hanna—have supported me personally and professionally throughout the decades. Thank you for accepting me as I am and somehow also pushing me to grow.

My loving parents, Jolanda and Willem, and incredible sister, Nina, and brother-in-law, Jason, have stopped at nothing to support my wildest dreams. You've kept me afloat, especially toward the end of this project. There aren't words to express how much you mean to me.

For over a decade, Robin has been my anchor. You've stood unwaveringly by my side since we were wide-eyed graduate students. Every triumph we've celebrated, including the life we're building together, I owe to you.

NOTES

CHAPTER 1: UNSILENCED SPRING

13 *more than 4.4 billion people live in cities:* World Bank Group. 2023. "Urban Development: Overview." World Bank. Accessed April 3, 2023. https://www .worldbank.org/en/topic/urbandevelopment/overview.

13 *a great suburbanization:* Andrew Stokols, "Suburbia Goes Global: What It Means for Urban Sustainability." TheCityFix, 2015. https://thecityfix.com /blog/suburbia-goes-global-what-it-means-urban-sustainability-beijing-china -andrew-stokols/.

14 *it's nearly doubled:* "Region of Waterloo." 2022. "Population." Accessed August 19, 2022. https://www.regionofwaterloo.ca/en/regional-government/population .aspx.

16 *80,000 km² of land:* World Bank Group. 2023. "Urban Development: Overview." World Bank. Accessed January 4, 2023. https://www.worldbank.org/en/topic /urbandevelopment/overview.

CHAPTER 2: SAVING TREES ON THE PAPER TRAIL

24 *wipe out 1.4 million urban trees by 2050:* Emily E. Puckett, Frank H. Koch, Mark J. Ambrose, and Brian Leung, "Hotspots of Pest-induced US Urban Tree Death, 2020–2050." *Journal of Applied Ecology* 60, no. 1, 2023, pp. 1–12. https:// besjournals.onlinelibrary.wiley.com/doi/abs/10.1111/1365-2664.14141.

24 *$6.8 billion in economic benefits to Americans:* David J. Nowak, Satoshi Hirabayashi, Allison Bodine, and Eric Greenfield. "Tree and Forest Effects on Air Quality and Human Health in the United States." *Environmental Pollution* 193 (2014): 119–129. https://www.sciencedirect.com/science/article/pii /S0269749114002395.

24 *same or better cooling benefits:* David J. Nowak, Robert E. Hoehn III, Daniel E. Crane, et al., "Assessing urban forest effects and values: Toronto's urban forest."

U.S. Department of Agriculture, Forest Service, Northern Research Station, 2007.

24 *up to 1,000 gallons of water annually:* "Urban Watershed Forestry Manual: Part 2: Conserving and Planting Trees at Development Sites." U.S. Department of Agriculture, Forest Service. Accessed September 20, 2022.

25 *calming effect on people:* Kirsten McEwan, Vanessa Potter, Yasuhiro Kotera, et al., "'This Is What the Colour Green Smells Like!': Urban Forest Bathing Improved Adolescent Nature Connection and Wellbeing." *International Journal of Environmental Research and Public Health* 19, no. 23 (2022): 15594.

25 *exposure to green spaces:* Hanneke Kruize, Nina van der Vliet, Brigit Staatsen, et al., "Urban green space: Creating a triple win for environmental sustainability, health, and health equity through behavior change." *International Journal of Environmental Research and Public Health* 16, no. 22 (2019): 4403.

25 *burgeoning field of research:* Rob McDonald, Timm Kroeger, Tim Boucher, et al., "Planting healthy air: A global analysis of the role of urban trees in addressing particulate matter pollution and extreme heat." *Planting Healthy Air,* The Nature Conservancy, 2016.

26 *didn't have a tree database:* "Tree inventory may ease storm reimbursements." *Sun-Sentinel,* September 23, 2010. Accessed November 20, 2021. https://www .sun-sentinel.com/2010/09/23/tree-inventory-may-ease-storm-reimbursements.

26 *U.S. cities without functional tree inventories:* Andrew K. Koeser, Richard J. Hauer, Robert J. Northrop, and Shawn M. Landry. "Municipal tree risk assessment in the United States: Findings from a comprehensive survey of urban forest management." *Arboricultural Journal* 38, no. 3 (2016): 175–192.

27 *Singapore is one of the world's greenest cities:* Cities Future. "How Singapore became one of the Greenest Cities in the World: 5 Key Reasons." Last modified 2023. Accessed May 12, 2023. https://citiesfuture.com/how-singapore-became -one-of-the-greenest-cities-in-the-world-5-key-reasons/.

27 *about 3,100 "tree incidents":* Oral Answer by Ministry of National Development on NParks' Current Tree Inspection Regime." Ministry of National Development, Singapore Government. Last modified 2023. Accessed March 2, 2023. https://www.mnd.gov.sg/newsroom/parliament-matters/q-as/view/oral -answer-by-ministry-of-national-development-on-nparks-current-tree-inspection -regime.

27 *nearly 90 percent decrease in annual tree incidents:* "Oral Answer by Ministry of National Development on NParks' Current Tree Inspection Regime." Ministry of National Development, Singapore Government. Last modified 2023. Accessed March 2, 2023. https://www.mnd.gov.sg/newsroom/parliament -matters/q-as/view/oral-answer-by-ministry-of-national-development-on -nparks-current-tree-inspection-regime.

29 *trees outnumbered the skyscrapers more than 500 to 1:* "New York City Trees Outnumber Skyscrapers 500 to 1." *Los Angeles Times.* September 22, 1996. Accessed February 15, 2023. https://www.latimes.com/archives/la-xpm-1996-09 -22-mn-46624-story.html.

30 *survivor of the attacks reflected:* Fire-Dex. (2012, September 11). The 9/11 Survivor Tree. Fire-Dex. https://blog.firedex.com/blog/2012/09/11/the-911-survivor-tree.

30 *"green accountant":* E. G. McPherson, "Accounting for benefits and costs of urban greenspace," *Landscape and Urban Planning* 22, no. 1 (1992), 41–51.

30 *Street Tree Resource Assessment Tool for Urban Forest Managers (STRATUM):* Greg McPherson, James R. Simpson, Paula J. Peper, et al., "Municipal Forest Benefits and Costs in Five US Cities," *Journal of Forestry* 103, no. 8 (2005): 411–416.

31 *West Coast tree species:* E. G. McPherson, James R. Simpson, Paula J. Peper, and Qingfu Xiao. "Benefit-cost analysis of Modesto's municipal urban forest," *Journal of Arboriculture* 25, no. 5 (1999): 235–248.

31 *trees provide $5.60 in benefits:* Paula J. Peper, E. Gregory McPherson, James R. Simpson, et al., "New York City, New York Municipal Forest Resource Analysis," *Center for Urban Forest Research, USDA Forest Service, Pacific Southwest Research Station, Davis* (2007).

32 *tree numbers had surged:* "New York City Parks, n.d. 'TreesCount! 2015.'"Accessed January 4, 2023. https://www.nyc.gov/parks/treemap.

34 *$1.5 billion set aside for the U.S. Forest Service Urban and Community Forestry program:* Jad Daley, "A New Day for Forestry," *American Forests.* Published July 28, 2022, last modified August 16. Accessed July 20, 2023. https://www.americanforests.org/blog/were-in-a-new-era-for-american-forestry/.

37 *35 percent of the city's trees stood on private property:* M. L. Treglia, M. Acosta-Morel, D. Crabtree, K. Galbo, et al., "The State of the Urban Forest in New York City." The Nature Conservancy, 2021. https://doi.org/10.5281/zenodo.5532876.

40 *a net loss of twenty-seven acres:* Grace Adams, "Outrage Over Illegal Tree Trimming Unlikely to Yield Prosecution," *Santa Monica Daily Press,* October 7, 2022. Accessed December 12, 2022. https://smdp.com/2022/10/07/outrage-over-illegal-tree-trimming-unlikely-to-yield-prosecution/.

42 *Canton ordered the company:* Ilya Shapiro, "My Oral Argument Debut Signals More Work to Be Done on Property Rights," Cato at Liberty, Cato Institute, November 2021.

42 *unconstitutional regulation:* Alan Greene, "Green Gold: Hands Off My Trees," Dykema, December 23, 2021. https://www.dykema.com/news-insights/green-gold-hands-off-my-trees.html.

CHAPTER 3: THROWING SHADE ON EXTREME HEAT

45 *suspended for more than ten days:* "Heatwave hits bridges, shipping in Amsterdam's waterways," DutchNews, August 2023. Accessed March 3, 2021. https://www.dutchnews.nl/2018/07/heatwave-hits-bridges-shipping-in-amsterdams-waterways/#:~:text=Several%20of%20Amsterdam%27s%20bridges%20are,tricky%20to%20open%20and%20close.

46 *one out of five Dutch people regularly ask older relatives:* "Temps up to 31 degrees today; Netherlands residents underestimate heat, says Red Cross," *NL Times,*

August 10, 2022. Accessed April 12, 2023. https://nltimes.nl/2022/08/10/temps-31-degrees-today-netherlands-residents-underestimate-heat-says-red-cross.

46   *Elderly women:* Yvette van Steen, Anna-Maria Ntarladima, Rick Grobbee, et al., "Sex differences in mortality after heat waves: Are elderly women at higher risk?," *International Archives of Occupational and Environmental Health* 92 (2019): 37–48.

46   *recorded 104,000 heat-related deaths among older people:* Will Bugler, "One hundred thousand deaths in a year: Europe tops mortality league for extreme heat." PreventionWeb, February 3, 2021. Accessed June 5, 2022. https://www.preventionweb.net/news/one-hundred-thousand-deaths-year-europe-tops-mortality-league-extreme-heat.

46   *July 2019:* Feng Ma, Xing Yuan, Yang Jiao, and Peng Ji. "Unprecedented Europe heat in June–July 2019: Risk in the historical and future context," *Geophysical Research Letters* 47, no. 11 (2020): e2020GL087809.

47   *2022, those heat waves were again surpassed:* Riyu Lu, Ke Xu, Ruidan Chen, et al., "Heat waves in summer 2022 and increasing concern regarding heat waves in general," *Atmospheric and Oceanic Science Letters* 16 (2023): 100290.

47   *Gallargues-le-Montueux rose to nearly 115°F:* Geeert Jan Van Oldenborgh, Sjoukje Philip, Sarah Kew, et al., "Human contribution to the record-breaking June 2019 heat wave in France," *World Weather Attribution* (2019).

47   *2003 European heat wave:* Dana Habeeb, Jason Vargo, and Brian Stone, "Rising heat wave trends in large US cities," *Natural Hazards* 76 (2015): 1651–1665.

47   *Europe's five hottest summers:* Andreas Hoy, Stephanie Hänsel, and Maurizio Maugeri, "An endless summer: 2018 heat episodes in Europe in the context of secular temperature variability and change," *International Journal of Climatology* 40, no. 15 (2020): 6315–6336.

47   *Earth's hottest month:* "Earth Just Had Its Hottest June on Record," National Oceanic and Atmospheric Administration, July 13, 2023. https://www.noaa.gov/news/earth-just-had-its-hottest-june-on-record.

47   *100 million people in the U.S. under excessive heat warnings:* "Heat Wave Live Updates: Warnings Issued for More than 100 Million Across the U.S.," NBC News, July 16, 2023. https://www.nbcnews.com/news/weather/live-blog/heat-wave-warnings-across-us-live-updates-rcna94490.

47   *Death Valley:* Seth Borenstein, John Locher, and Adam Beam. "California's Death Valley Sizzles as Brutal Heat Wave Continues," AP News, July 16, 2023. https://apnews.com/article/death-valley-heat-wave-california-hottest-record-c1b2d83dc384e46f133d460893787c52.

47   *China registered record-breaking hot days:* "China Logs 52.2 Celsius as Extreme Weather Rewrites Records." *Reuters,* July 17, 2023, sec. China. https://www.reuters.com/world/china/china-logs-522-celsius-extreme-weather-rewrites-records-2023-07-17/.

47   *road sign worker:* Julia Malleck, "The Death of a Road Worker During Italy's Hellish Heat Wave Is a Labor Rights Issue," Quartz, July 13, 2023. https://qz.com/italy-cerberus-heat-wave-death-road-worker-labor-rights-1850637007.

47   *Acropolis:* "Greece Briefly Shuts Acropolis Site to Protect Tourists from Heatwave." *Reuters,* July 14, 2023, sec. Europe. https://www.reuters.com /world/europe/greece-briefly-shuts-acropolis-site-protect-tourists-heatwave -2023-07-14/.

47   *evacuate resort towns:* Helena Smith, "Greece 'at War with Fire' Amid Chaotic Evacuation of Tourists from Rhodes," *The Guardian,* July 24, 2023, sec. World News. https://www.theguardian.com/world/2023/jul/24/greece -wildfires-corfu-evia-rhodes-heatwave-northern-hemisphere-extreme-weather -temperatures-europe.

48   *Europe's inreasingly fatal summers:* Marcel Alied and Nguyen Tien Huy, "A reminder to keep an eye on older people during heatwaves," *Lancet Healthy Longevity* 3, no. 10 (2022): e647–e648.

48   *environmental determinants of health:* Kim R. van Daalen, Marina Romanello, Joacim Rocklöv, et al., "The 2022 Europe report of the Lancet Countdown on health and climate change: Towards a climate resilient future," *Lancet Public Health* 7, no. 11 (2022): e942–e965.

48   *little reason to design for them:* Quirin Schiermeier, "Climate change made Europe's mega-heatwave five times more likely," *Nature* 571, no. 7764 (2019): 155–156.

48   *seasonal affective disorder:* Barbara Nussbaumer-Streit, D. Winkler, M. Spies, et al., "Prevention of seasonal affective disorder: Results of a survey in German-speaking countries: Barbara Nussbaumer-Streit." *European Journal of Public Health* 27, no. suppl_3 (2017): ckx189–178.

49   *fires in coastal Attica:* Stathis G. Arapostathis, "Instagrammers report about the deadly wildfires of East Attica, 2018, Greece: An introductory analytic assessment for disaster management purposes," *Proceedings of the 16th International Conference on Information Systems for Crisis Response and Management,* ISCRAM, 2019.

49   *eight hundred grass and open-land blazes:* "London Heatwave: More than 800 Grass Fires in Six Weeks," BBC News, July 16, 2022, sec. London. https://www .bbc.com/news/uk-england-london-62190509.

49   *brigade's busiest day since World War II:* "London Fire Brigade Had Busiest Day Since World War Two, Says London Mayor," BBC News, July 20, 2022, sec. UK. https://www.bbc.com/news/uk-62232654.

49   *losing 36 million trees:* David J. Nowak and Eric J. Greenfield. "Declining urban and community tree cover in the United States," *Urban Forestry & Urban Greening* 32 (2018): 32–55.

49   *lower summer daytime temperatures:* "Using trees and vegetation to reduce heat islands," Heat Islands, U.S. Environmental Protection Agency, 2016.

49   *additional 1,200 heat-related deaths annually:* Robert I. McDonald, Timm Kroeger, Ping Zhang, and Perrine Hamel, "The value of US urban tree cover for reducing heat-related health impacts and electricity consumption," *Ecosystems* 23 (2020): 137–150.

50   *"heat dome":* Joanne Silberner, "Heat wave causes hundreds of deaths and hospitalisations in Pacific north west." \ *British Medical Journal (Online)* 374 (2021).

50   *urban heat island effect:* Laura Kleerekoper, Marjolein Van Esch, and Tadeo
      Baldiri Salcedo, "How to make a city climate-proof, addressing the urban heat
      island effect," *Resources, Conservation and Recycling* 64 (2012): 30–38.

50   *shade and evapotranspiration:* Alexander Thomas Hayes, Zahra Jandaghian, Mi-
      chael A. Lacasse, et al., "Nature-based solutions (nbss) to mitigate urban heat
      island (UHI) effects in Canadian cities," *Buildings* 12, no. 7 (2022): 925.

51   *significantly reduce the urban health island effect:* Yujia Zhang, Alan T. Murray,
      and B. L. Turner Ii, "Optimizing green space locations to reduce daytime and
      nighttime urban heat island effects in Phoenix, Arizona," *Landscape and Urban
      Planning* 165 (2017): 162–171.

53   *studying heat across urban neighborhoods:* Vivek Shanda and Yasuyo Makido,
      "Cooling the City: Integrating ground-based measurements with modeling
      scenarios to address urban heat stress among vulnerable populations," (2017).
      https://www2.jpgu.org/meeting/2017/PDF2017/H-DS10_all.pdf.

54   *highest death toll:* Jacklyn N. Kohon, Katsuya Tanaka, Dani Himes, et al., "Ex-
      treme Heat Vulnerability among Older Adults: A Multi-level Risk Index for
      Portland, Oregon." *The Gerontologist* (2023): gnad074.

55   *Decades of heat research:* Jeremy S. Hoffman, Vivek Shandas, and Nicholas Pen-
      dleton, "The effects of historical housing policies on resident exposure to intra-
      urban heat: A study of 108 US urban areas," *Climate* 8, no. 1 (2020): 12.

57   *Smaller plants also evapotranspire:* Stefano Cascone, Julià Coma, Antonio
      Gagliano, and Gabriel Pérez, "The evapotranspiration process in green roofs: A
      review," *Building and Environment* 147 (2019): 337–355.

58   *ivy to be the most effective plant cover:* Faye Thomsit-Ireland, Emmanuel A. Essah,
      Paul Hadley, and Tijana Blanuša, "The impact of green facades and vegetative
      cover on the temperature and relative humidity within model buildings," *Build-
      ing and Environment* 181 (2020): 107009.

58   *30 percent less heat loss:* Matthew Fox, Jack Morewood, Thomas Murphy, Paul
      Lunt, and Steve Goodhew. "Living wall systems for improved thermal perfor-
      mance of existing buildings." *Building and Environment* 207 (2022): 108491.

58   *people paying attention*: Mike Baker and Sergio Olmos. "The Pacific Northwest,
      Built for Mild Summers, Is Scorching Yet Again," *New York Times*, August 13,
      2021, sec. U.S. https://www.nytimes.com/2021/08/13/us/excessive-heat-warning
      -seattle-portland.html.

58   *the vulnerabilities of Portland:* Jackson Voelkel, Dana Hellman, Ryu Sakuma,
      and Vivek Shandas, "Assessing vulnerability to urban heat: A study of dispro-
      portionate heat exposure and access to refuge by socio-demographic status in
      Portland, Oregon," *International Journal of Environmental Research and Public
      Health* 15, no. 4 (2018): 640.

59   *land-surface temperature:* Vivek Shandas, Jackson Voelkel, Joseph Williams, and
      Jeremy Hoffman, "Integrating satellite and ground measurements for predicting
      locations of extreme urban heat," *Climate* 7, no. 1 (2019): 5.

59   *measurements are coarse:* Kevan B. Moffett, Yasuyo Makido, and Vivek Shan-

das, "Urban-rural surface temperature deviation and intra-urban variations contained by an urban growth boundary," *Remote Sensing* 11, no. 22 (2019): 2683.

59   *raise the risk of dying by 2.5 percent:* G. Brooke Anderson and Michelle L. Bell, "Heat waves in the United States: Mortality risk during heat waves and effect modification by heat wave characteristics in 43 US communities," *Environmental Health Perspectives* 119, no. 2 (2011): 210–218.

60   *twice as many deaths:* Justine Hausheer, "Trees in the US Annually Prevent 1,200 Deaths during Heat Waves," Cool Green Science, May 8, 2019. https://blog .nature.org/2019/05/08/trees-in-the-us-annually-prevent-1200-deaths-during -heat-waves/#:~:text=In%20these%2097%20cities%2C%20trees.

61   *another name for it: Treepedia:* Bill Yang Cai, Xiaojiang Li, Ian Seiferling, and Carlo Ratti, "Treepedia 2.0: Applying deep learning for large-scale quantification of urban tree cover," *2018 IEEE International Congress on Big Data (Big-Data Congress),* pp. 49–56.

62   *maximizing cooling benefits: Hannah* Furfaro, "New tool lets cities see where trees are needed," *Wall Street Journal,* April 16, 2017.

62   *technology that showed indisputable proof:* Sinéad Nicholson, Marika Tomasi, Daniele Belleri, et al., "'Greening' the Cities: How Data Can Drive Interdisciplinary Connections to Foster Ecological Solutions." *SPOOL* 9, no. 1 (2022): 5–18.

63   *community-based heat mapping campaign:* "CAPA Strategies." Accessed June 25, 2023. https://www.capastrategies.com.

63   *13 percent rise in anxiety over the effects of climate change:* Andreea Bratu, Kiffer G. Card, Kalysha Closson, et al., "The 2021 Western North American heat dome increased climate change anxiety among British Columbians: Results from a natural experiment," *Journal of Climate Change and Health* 6 (2022): 100116.

63   *Durham and Raleigh:* "Urban Heat Island Temperature Mapping Campaign— North Carolina State Climate Office," Urban Heat Island Temperature Mapping Campaign, March 18, 2021. https://climate.ncsu.edu/research/uhi/.

64   *dramatic temperature differences:* Bradley Wilson, Jeremy R. Porter, Edward J. Kearns, et al., "High-Resolution Estimation of Monthly Air Temperature from Joint Modeling of In Situ Measurements and Gridded Temperature Data," *Climate* 10, no. 3 (2022): 47.

65   *thirty-plus cities on Treepedia:* "Treepedia: MIT Senseable City Lab," 2023. https://senseable.mit.edu/treepedia.

65   *green space per inhabitant:* "Who Benefits from Nature in Cities? Social Inequalities in Access to Urban Green and Blue Spaces across Europe—European Environment Agency," European Environment Agency, February 7, 2023. https:// www.eea.europa.eu/publications/who-benefits-from-nature-in/who-benefits -from-nature-in.

65   *Mayor Hidalgo:* India Block, "Paris Plans to Green by Planting 'Urban Forest' around Architectural Landmarks," Dezeen, June 26, 2019. https://www.dezeen .com/2019/06/26/paris-urban-forest-plant-trees-landmarks/.

CHAPTER 4: SPREADING LIKE WILDFIRE

67   *Eighty-five lives were cut short:* Priyanka Boghani, "Camp Fire: By the Numbers," *Frontline*, PBS, October 29, 2019. Accessed July 9, 2022. https://www.pbs.org/wgbh/frontline/article/camp-fire-by-the-numbers/.

68   *had only gotten one thirty-fifth of that:* Joseph Serna and Rong-Gong Lin II, "First rain in months douses California wildfire but raises the risk of mudslides," *Los Angeles Times,* November 21, 2018. Accessed March 30, 2022. https://www.latimes.com/local/lanow/la-me-rain-fires-california-20181113-story.html.

69   *throwing sparks into the dry brush:* Peter Eavis and Ivan Penn, "California Says PG&E Power Lines Caused Camp Fire That Killed 85," *New York Times,* May 15, 2019. Accessed April 1, 2021. https://www.nytimes.com/2019/05/15/business/pge-fire.html.

69   *6.6 percent burned from 2002 to 2011:* "California forests hit hard by wildfires in the last decade," Wildfire Today, September 5, 2022. Accessed October 25, 2022. https://wildfiretoday.com/2022/09/05/california-forests-hit-hard-by-wildfires-in-the-last-decade/.

69   *destroyed at least 5,600 homes and structures:* "2017 Tubbs Fire Incident Information," CAL FIRE, October 8, 2017. Accessed January 8, 2023. https://www.fire.ca.gov/incidents/2017/10/8/tubbs-fire-central-lnu-complex/.

70   *PG&E have caused over 1,500 wildfires:* Russell Gold, Katherine Blunt, and Rebecca Smith, "PG&E Sparked at Least 1,500 California Fires. Now the Utility Faces Collapse." *Wall Street Journal*, January 13, 2019. Accessed July 20, 2023. https://www.wsj.com/articles/pg-e-sparked-at-least-1-500-california-fires-now-the-utility-faces-collapse-11547410768.

70   *the "gigafire":* "California fire is now a 'gigafire,' a rare designation," CNN, October 6, 2020. Accessed December 21, 2022. https://www.cnn.com/2020/10/06/us/gigafire-california-august-complex-trnd/index.html.

71   *15,000 separate fires:* Brendan D. Cowled, Melanie Bannister-Tyrrell, Mark Doyle, et al., "The Australian 2019/2020 black summer bushfires: Analysis of the pathology, treatment strategies and decision making about burnt livestock," *Frontiers in Veterinary Science* 9 (2022): 790556.

71   *damaged the ozone layer:* Lilly Damany-Pearce, Ben Johnson, Alice Wells, et al., "Australian wildfires cause the largest stratospheric warming since Pinatubo and extends the lifetime of the Antarctic ozone hole," *Scientific Reports* 12, no. 1 (2022): 12665.

71   *killing at least eighty-six:* J. San-Miguel-Ayanz, T. Durrant, R. Boca, et al., "Forest Fires in Europe, Middle East and North Africa 2021," EUR 31269 EN, Publications Office of the European Union, Luxembourg, 2022, ISBN 978-92-76-58585-5, doi:10.2760/34094, JRC130846.

71   *eight of the most extreme wildfire years have occured in the last decade:* Piyush Jain, Dante Castellanos-Acuna, Sean CP Coogan, et al., "Observed increases in extreme fire weather driven by atmospheric humidity and temperature," *Nature Climate Change* 12, no. 1 (2022): 63–70.

72 *forced to move:* Mahalia B. Clark, Ephraim Nkonya, and Gillian L. Galford, "Flocking to fire: How climate and natural hazards shape human migration across the United States," *Frontiers in Human Dynamics* 4 (2022): 46.

72 *federal disaster funds:* Claudia Boyd-Barrett, "As Wildfires Grow, So Does California's Housing and Homelessness Crisis. Here Are Some Solutions," *California Health Report,* August 3, 2022. Accessed January 13, 2023. https://www.calhealthreport.org/2022/08/03/as-wildfires-grow-so-does-californias-housing-and-homelessness-crisis-here-are-some-solutions/.

78 *Adverse Childhood Experiences:* "Butte Thrives- Coalition on Adverse Childhood Experiences," HelpCentral.org, Connecting Butte County with Free and Low-Cost Services, April 11, 2018. Accessed December 1, 2022. http://helpcentral.org/buttethrives/.

79 *approximately 80,000 acres annually:* Bill Gabbert, "California Agencies Intend to Ramp Up Prescribed Burning," Wildfire Today, April 4, 2022. Accessed May 21, 2023. https://wildfiretoday.com/2022/04/04/california-agencies-intend-to-ramp-up-prescribed-burning/.

80 *two times as much greenhouse gas emissions:* Michael Jerrett, Amir S. Jina, and Miriam E. Marlier, "Up in smoke: California's greenhouse gas reductions could be wiped out by 2020 wildfires," *Environmental Pollution* 310 (2022): 119888.

84 twice *as toxic:* Athanasios Nenes, "'Four times more toxic': How wildfire smoke ages over time," Horizon: The EU Research & Innovation Magazine, July 20, 2020. Accessed March 21, 2021. https://ec.europa.eu/research-and-innovation/en/horizon-magazine/four-times-more-toxic-how-wildfire-smoke-ages-over-time.

84 *the most land burned in any year:* Dion, Mathieu. "Hundreds of Fires Are Out of Control in Canada's Worst-Ever Season." Bloomberg News, June 7, 2023. Accessed August 27, 2023. https://www.bloomberg.com/news/articles/2023-06-07/hundreds-of-fires-are-out-of-control-in-canada-s-worst-ever-season.

84 *120 million Americans:* Salahieh, Nouran, Joe Sutton, and Lauren Mascarenhas. "Smoke from Canada's wildfires is drifting across the Great Lakes." CNN, June 28, 2023. Accessed September 5, 2023. https://edition.cnn.com/2023/06/27/us/canada-wildfire-smoke-great-lakes/index.html.

84 *Air Quality Index:* "Home." NC State University Air Quality Climate Group. Accessed September 1, 2023. https://airquality.climate.ncsu.edu/.

84 *IQAir:* IQAir. "New York City Air Quality Index (AQI) and New York Pollution | AirVisual." Accessed September 8, 2023. https://www.iqair.com/us/usa/new-york/new-york-city.

86 *380,000-volt electric cables underground:* DutchNews.nl. "High Voltage Power Cables to Go Underground," DutchNews.nl, January 12, 2009. https://www.dutchnews.nl/2009/01/high_voltage_power_cables_to_g/.

88 *increase its high-voltage transmission lines:* Andrew Pascale and Eric D. Larson, "Net Zero America Study," Princeton University. Accessed March 23, 2023.

https://netzeroamerica.princeton.edu/img/Princeton%20NZA%20FINAL
%20REPORT%20SUMMARY%20(29Oct2021).pdf.

89 *pointing a finger:* Associated Press. "Hawaiian Electric Company: who is responsible for igniting wildfire in Maui?" *The Guardian,* August 29, 2023. Accessed
September 3, 2023. https://www.theguardian.com/us-news/2023/aug/28/maui
-wildfires-hawaiian-electric-company

90 *minimal progress:* Ailworth, Erin. "Maui's Fires and the Electric Grid." *Wall
Street Journal.* August 6, 2023. Accessed September 1, 2023. https://www.wsj
.com/articles/maui-fires-electric-grid-hawaiian-electric-green-energy-2b2c1399.

90 *mandatory push:* Blunt, Katherine, Dan Frosch, and Jim Carlton. "Wildfire
Risk Rises on Hawaii's Maui as Hawaiian Electric Shuts Off Power." *Wall Street
Journal.* August 17, 2023. Accessed September 6, 2023. https://www.wsj.com
/us-news/wildfire-risk-maui-hawaiian-electric-7beed21e.

90 *fully renewable grid by 2045:* Hawaii State Energy Office. "Clean Energy Vision."
Accessed September 6, 2023. https://energy.hawaii.gov/what-we-do/clean
-energy-vision/.

90 *regulatory framework:* The Daily Wire. "Hawaii Utility Pursued Green Energy
Goals While Fire Mitigation Projects Were Delayed." Accessed September 2,
2023. https://www.dailywire.com/news/hawaii-utility-pursued-green-energy
-goals-while-fire-mitigation-projects-were-delayed.

99 *Walbridge Fire:* Incidents | CAL FIRE—CA.gov. Fire.ca.gov. Accessed June 3,
2023.https://www.fire.ca.gov/incidents.

100 *CodeRED:* Simon Romero, Tim Arango, and Thomas Fuller, "A Frantic Call,
a Neighbor's Knock, but Few Official Alerts as Wildfire Closed In," *New York
Times,* November 21, 2018. Accessed April 28, 2023. https://www.nytimes
.com/2018/11/21/us/paradise-fires-emergency-alerts.html.

100 *specializes in emergency management:* Erica D. Kuligowski, Nicholas A. Waugh,
Jeannette Sutton, and Thomas J. Cova, "Ember Alerts: Assessing Wireless Emergency Alert Messages in Wildfires Using the Warning Response Model," *Natural Hazards Review* 24, no. 2 (2023): 04023009.

101 *alerting the public to natural disasters:* Hawaii Emergency Management Agency.
"All Hazard Statewide Outdoor Warning Siren System." Accessed September 24, 2023. URL: https://dod.hawaii.gov/hiema/all-hazard-statewide-outdoor
-warning-siren-system/.

102 *code of conduct:* "Code of Conduct," Watch Duty. Accessed December 22, 2023.
https://www.watchduty.org/how-it-works/code-of-conduct.

CHAPTER 5: DRAINING THE SWAMP IN REAL TIME

106 *left 90 percent of the village . . . in ruins:* Amy Smart, "Trial by fire continues for
Lytton, B.C., its residents in limbo, buildings in ruins," *Canadian Press,* June 24,
2022. Accessed July 23, 2023. https://www.theglobeandmail.com/canada
/british-columbia/article-trial-by-fire-continues-for-lytton-bc-its-residents-in
-limbo-buildings/.

107 *will not be repaired until 2024:* Evan Saunders, "B.C. Ministry of Transportation and Infrastructure outlines more than \$12 billion of 2023 priorities," *Journal of Commerce,* February 13, 2023. Accessed March 13, 2023. https://canada .constructconnect.com/joc/news/infrastructure/2023/02/b-c-department-of -infrastructure-outlines-more-than-12-billion-of-2023-priorities.

111 *entire villages were swept away:* Herman Gerritsen, "What happened in 1953? The Big Flood in the Netherlands in retrospect," *Philosophical Transactions of the Royal Society A: Mathematical, Physical and Engineering Sciences* 363, no. 1831 (2005): 1271–1291.

111 *"Seven Wonders of the Modern World":* "Seven Wonders," American Society of Civil Engineers, n.d. Accessed July 23, 2023. https://www.asce.org/seven -wonders/.

113 *officials were later arrested for deliberately underreporting or concealing:* Helen Davidson, "Chinese officials arrested for concealing true scale of flood death toll," *The Guardian,* January 2, 2022. Accessed July 23, 2023. https://www .theguardian.com/world/2022/jan/23/chinese-provincial-officials-concealed -scores-of-deaths-from-flood-disaster.

114 *outdated CSS:* John Tibbetts, "Combined sewer systems: Down, dirty, and out of date," *Environmental Health Perspectives* 113, no. 7 (2005): A464–A467.

114 *concentrated in the Northeast and Great Lakes regions:* John Tibbetts, "Combined sewer systems: Down, dirty, and out of date," *Environmental Health Perspectives* 113, no. 7 (2005): A464–A467.

114 *roughly half of all sewer systems are combined:* Emanuele Quaranta and Alberto Pistocchi, "Combined Sewer Overflows: Impacts, mitigation solutions and European perspective," Slow The Flow, September 24, 2022. Accessed May 23, 2023. https://slowtheflow.net/combined-sewer-overflows-impacts-mitigation -solutions-and-european-perspective/.

114 *80 percent of residents . . . rely on CSSs:* Hisashi Tsutsui and Kim Chuni, "Tokyo Bay sewage raises Olympian stink at 2020 swimming site," *Nikkei Asia,* September 26, 2019. Accessed August 1, 2021. https://asia.nikkei.com/Spotlight /Environment/Tokyo-Bay-sewage-raises-Olympian-stink-at-2020-swimming -site.

114 *470,000 tons daily into the Liuxi River:* Danielle Neighbour and Gillian Zwicker, "What China Can Learn from New York City about Wastewater Management," *Scientific American,* March 8, 2019. Accessed April 7, 2020. https://blogs .scientificamerican.com/observations/what-china-can-learn-from-new-york -city-about-wastewater-management/.

115 *CSOs release around 850 billion gallons of diluted yet untreated sewage:* "Clean Water Act: EPA Should Track Control of Combined Sewer Overflows and Water Quality Improvements," United States Government Accountability Office, January 2023. Accessed July 23, 2023. https://www.gao.gov/assets/gao -23-05285.pdf.

115 *\$50 million on green infrastructure:* "Chicago Announces First Projects Under Green Infrastructure Strategy," Stormwater Report, May 15, 2014. Accessed

June 23, 2022. https://stormwater.wef.org/2014/05/chicago-announces-first
-projects-green-infrastructure-strategy/.

116 *total estimated cost of $735 million:* Michael Hawthorne, "Flooding in the
Chicago area has been so bad in the past decade that only places ravaged by
hurricanes sustain more damage," *Chicago Tribune*, May 10, 2019. Accessed August 5, 2022. https://www.chicagotribune.com/news/breaking/ct-met-chicago
-flooding-basement-sewage-20190506-story.html.

118 *Digital Water City (DWC):* Mathias Riechel, Oriol Gutierrez, Silvia Busquets,
et al., "A network of low-cost temperature sensors for real-time monitoring of
combined sewer overflow." In *EGU General Assembly Conference Abstracts*, 2021,
EGU21–16386.

120 *halt a fifth of combined overflow:* Emanuele Quaranta, Stephan Fuchs, Hendrik
Jan Liefting, Alma Schellart, and Alberto Pistocchi, "Costs and benefits of combined sewer overflow management strategies at the European scale," *Journal of
Environmental Management* 318 (2022): 115629.

122 *sustained up to $19 billion in damages:* Jeroen C.J.H. Aerts, Wouter Botzen, Hans
de Moel, and Kerry Emanuel, "Hurricane Sandy and adaptation pathways in
New York: Lessons from a first-responder city," *Global Environmental Change*
28 (2014): 395–408. Accessed December 13, 2022. https://www.sciencedirect
.com/science/article/pii/S0959378014000910/.

126 *obliteration of a much-loved . . . waterfront park:* Nathan Kensinger, "NYC Says
Goodbye to East River Park," *Gothamist*, December 31, 2021. Accessed September 24, 2022. https://gothamist.com/news/photos-nyc-says-goodbye-east
-river-park.

128 *rules:* Emma Shaw, "14 Crazy Singapore Laws You Should Know before You
Go," Explore ShawTravel Blog, February 27, 2018. https://www.exploreshaw
.com/14-crazy-singapore-laws-to-know-before/.

129 *top one hundred places:* Goody Feed Team, "100 S'pore Flood-Prone Areas
You Must Avoid When It's Raining Heavily" Goody Feed, December 11, 2015.
https://goodyfeed.com/100-spore-flood-prone-areas-you-must-avoid-when
-its-raining-heavily/.

130 *growing network of sensors:* "'Google Drains'—Changi Airport's New Layer
of Defence Against Flash Floods," n.d. Changiairport.com. https://www
.changiairport.com/corporate/media-centre/changijourneys/the-airport
-never-sleeps/google-drain-map.html.

CHAPTER 6: THE SCIENTISTS WITHIN US

134 *ignorance about the natural world:* Yves Zinngrebe, "Learning from local knowledge in Peru—Ideas for more effective biodiversity conservation," *Journal for
Nature Conservation* 32 (2016): 10–21.

135 *thousands of species going extinct:* Jeremy Hance, "Brazil's Atlantic Forest
(Mata Atlântica)," Mongabay, n.d. https://rainforests.mongabay.com/mata
-atlantica/.

135   *disrupt natural ecosystems:* Ngozi H. Arihilalm and E. C. Arihilam, "Impact and control of anthropogenic pollution on the ecosystem—A review," *Journal of Bioscience and Biotechnology Discovery* 4, no. 3 (2019): 54–59.

136   *cities with a wide range of plant and animal species:* Cathy Oke, Sarah A. Bekessy, Niki Frantzeskaki, et al., "Cities should respond to the biodiversity extinction crisis," *npj Urban Sustainability* 1, no. 1 (2021): 11.

136   *pollination plays a vital role:* Callum J. Macgregor and Alison S. Scott-Brown, "Nocturnal pollination: An overlooked ecosystem service vulnerable to environmental change," *Emerging Topics in Life Sciences* 4, no. 1 (2020): 19–32.

136   *biodiversity can protect us:* Sajjad Ali, Muhammad Irfan Ullah, Asif Sajjad, et al., "Environmental and health effects of pesticide residues," *Sustainable Agriculture Reviews 48: Pesticide Occurrence, Analysis and Remediation, Vol. 2 Analysis* (Springer 2020), 311–336.

137   *"plant blindness":* Alexandros Amprazis and Penelope Papadopoulou, "Plant blindness: A faddish research interest or a substantive impediment to achieve sustainable development goals?," *Environmental Education Research* 26, no. 8 (2020): 1065–1087.

137   *"urban evolution":* Menno Schilthuizen, *Darwin comes to town: How the urban jungle drives evolution* (Picador, 2019).

137   *birds have pitched their songs higher:* David A. Luther, Jennifer Phillips, and Elizabeth P. Derryberry, "Not so sexy in the city: Urban birds adjust songs to noise but compromise vocal performance," *Behavioral Ecology* 27, no. 1 (2016): 332–340.

137   *digest fatty foods:* Menno Schilthuizen, *Darwin comes to town: How the urban jungle drives evolution* (Picador, 2019).

138   *model species:* Niels A.G. Kerstes, Thijmen Breeschoten, Vincent J. Kalkman, and Menno Schilthuizen, "Snail shell colour evolution in urban heat islands detected via citizen science," *Communications Biology* 2, no. 1 (2019): 264.

138   *"super-coyotes":* Alexandra L. DeCandia, Carol S. Henger, Amelia Krause, et al., "Genetics of urban colonization: Neutral and adaptive variation in coyotes (Canis latrans) inhabiting the New York metropolitan area," *Journal of Urban Ecology* 5, no. 1 (2019): juz002.

138   *attacking full-grown adults:* Lynsey A. White and Stanley D. Gehrt, "Coyote attacks on humans in the United States and Canada," *Human Dimensions of Wildlife* 14, no. 6 (2009): 419–432.

139   *garden snail's shell results from its DNA:* Niels A.G. Kerstes, Thijmen Breeschoten, Vincent J. Kalkman, and Menno Schilthuizen. "Snail shell colour evolution in urban heat islands detected via citizen science," *Communications Biology* 2, no. 1 (2019): 264.

139   *temperature differences:* Timothy J. Bauer, "Interaction of urban heat island effects and land-sea breezes during a New York City heat event," *Journal of Applied Meteorology and Climatology* 59, no. 3 (2020): 477–495.

140   *only 1.5 million species have been named:* Iva Njunjić, "How I transformed Leonardo DiCaprio into a beetle," TEDxGhent. Filmed July 2019. YouTube video,

10:57. Accessed February 2, 2023. https://www.ted.com/talks/iva_njunjic_how
_i_transformed_leonardo_dicaprio_into_a_beetle?language=en.

140 *only one scientist for every fifteen hundred species of insects:* Iva Njunjić, "How
I transformed Leonardo DiCaprio into a beetle," TEDxGhent. Filmed July
2019. YouTube video, 10:57. Accessed February 2, 2023. https://www.ted.com
/talks/iva_njunjic_how_i_transformed_leonardo_dicaprio_into_a_beetle
?language=en.

140 *dropped 76 percent since 1989:* Caspar A. Hallmann, Martin Sorg, Eelke Jongejans, et al., "More than 75 percent decline over 27 years in total flying insect
biomass in protected areas," *PLoS ONE 12*, no. 10 (2017): e0185809. Accessed
July 23, 2023. https://doi.org/10.1371/journal.pone.0185809.

140 *inspecting windshield splatter:* Anders Pape Møller, "Parallel declines in abundance of insects and insectivorous birds in Denmark over 22 years," *Ecology and
Evolution* 9, no. 11 (2019): 6581–6587.

140 *insects are declining at nine times the rate:* Francisco Sánchez-Bayo and Kris A.
Wyckhuys, "Worldwide decline of the entomofauna: A review of its drivers,"
Biological Conservation 232, (2019): 8–27. Accessed July 23, 2023. https://doi
.org/10.1016/j.biocon.2019.01.020.

141 *insect populations have diminished by 45 percent worldwide:* Pedro Cardoso,
Philip S. Barton, Klaus Birkhofer, et al., "Scientists' warning to humanity on
insect extinctions," *Biological Conservation 242,* (2020): 108426. Accessed
July 23, 2023. https://doi.org/10.1016/j.biocon.2020.108426.

141 *wiping out species faster:* Rodolfo Dirzo, Hillary S. Young, Mauro Galetti, et al.,
"Defaunation in the Anthropocene," *Science* 345, no. 6195 (2014): 401–406.

142 *disrupt the behavior:* Avalon C.S. Owens and Sara M. Lewis, "Artificial light impacts the mate success of female fireflies," *Royal Society Open Science* 9, no. 8
(2022): 220468.

143 *keystone species:* Luciana Elizalde, Marina Arbetman, Xavier Arnan, et al., "The
ecosystem services provided by social insects: Traits, management tools and
knowledge gaps," *Biological Reviews* 95, no. 5 (2020): 1418–1441.

145 *observation.org:* Observation.org., 2020. https://observation.org/.

146 *"ecological Armageddon":* Caspar A. Hallmann, Martin Sorg, Eelke Jongejans,
et al., "More than 75 percent decline over 27 years in total flying insect biomass
in protected areas," *PloS One* 12, no. 10 (2017): e0185809.

146 *insect traps and weighing techniques:* Graham A. Montgomery, Michael W. Be-
litz, Rob P. Guralnick, and Morgan W. Tingley, "Standards and best practices
for monitoring and benchmarking insects," *Frontiers in Ecology and Evolution* 8
(2021): 513

146 *challenging research area:* Roel van Klink, Tom August, Yves Bas, et al.,
"Emerging technologies revolutionise insect ecology and monitoring," *Trends
in Ecology & Evolution* 37, no. 10 (2022): 872–885.

147 *DIOPSIS insect camera:* Roel van Klink, Tom August, Yves Bas, et al., "Emerging technologies revolutionise insect ecology and monitoring," *Trends in Ecology & Evolution* 37, no. 10 (2022): 872–885.

150   *partnership with Snapchat:* "PlantSnap Brings Plant Recognition to Snapchat," PlantSnap Inc. 2020. Accessed August 3, 2021. https://www.prnewswire.com /ae/news-releases/plantsnap-brings-plant-recognition-to-snapchat-301074631 .html#:~:text=TELLURIDE%2C%20Colo.%2C%20June%2011.

151   "biodiversity big data": Ellinore J. Theobald, Ailene K. Ettinger, Hillary K. Burgess, et al., "Global change and local solutions: Tapping the unrealized potential of citizen science for biodiversity research," *Biological Conservation* 181 (2015): 236–244.

151   *impacts of the users' observations:* Alison Johnston, Eleni Matechou, and Emily B. Dennis, "Outstanding challenges and future directions for biodiversity monitoring using citizen science data," *Methods in Ecology and Evolution* 14, no. 1 (2023): 103–116.

151   *citizen-science revolution:* Corey T. Callaghan, Alistair G.B. Poore, Thomas Mesaglio, et al., "Three frontiers for the future of biodiversity research using citizen science data," *BioScience* 71, no. 1 (2021): 55–63.

151   *Democratizing biodiversity data collection:* J. Mason Heberling, Joseph T. Miller, Daniel Noesgaard, et al., "Data integration enables global biodiversity synthesis," *Proceedings of the National Academy of Sciences* 118, no. 6 (2021): e2018093118.

151   *citizen science has yet to fill the gap:* Anna K. Monfils, Erica R. Krimmel, John M. Bates, et al., "Regional collections are an essential component of biodiversity research infrastructure," *BioScience* 70, no. 12 (2020): 1045–1047.

151   *smartphone-addicted people:* Jay A. Olson, Dasha A. Sandra, Élissa S. Colucci, et al., "Smartphone addiction is increasing across the world: A meta-analysis of 24 countries," *Computers in Human Behavior 129,* (2022): 107138. Accessed March 15, 2023. https://doi.org/10.1016/j.chb.2021.107138.

152   *benefits of citizen science:* Michael J.O. Pocock, Iain Hamlin, Jennifer Christelow, Holli-Anne Passmore, and Miles Richardson, "The benefits of citizen science and nature-noticing activities for well-being, nature connectedness and pro-nature conservation behaviours," *People and Nature* 5, no. 2 (2023): 591–606.

152   *citizen-science activities:* Helen E. Roy, Michael J.O. Pocock, Chris D. Preston, et al., "Understanding citizen science and environmental monitoring: Final report on behalf of UK Environmental Observation Framework," NERC Centre for Ecology & Hydrology and Natural History Museum (2012).

152   *Gen Z:* Rory Burke, Orla L. Sherwood, Stephanie Clune, et al., "Botanical boom: A new opportunity to promote the public appreciation of botany," *Plants, People, Planet* 4, no. 4 (2022): 326–334.

152   *plant blindness as a major contributor to obesity, mental illness:* Sarah Elton, "Relational health: Theorizing plants as health-supporting actors," *Social Science & Medicine* 281 (2021): 114083.

153   *harnessed the collective expertise of amateur birders worldwide:* Robert Stevenson, Carl Merrill, and Peter Burn, "Useful biodiversity data were obtained by novice observers using iNaturalist during college orientation retreats," *Citizen Science: Theory and Practice* 6, no. 1 (2021).

153   *eBird:* Connor J. Rosenblatt, Ashley A. Dayer, Jennifer N. Duberstein, et al.,

"Highly specialized recreationists contribute the most to the citizen science project eBird," *Ornithological Applications* 124, no. 2 (2022): duac008.

153   *birdsong-recognition apps:* Benjamin M. Van Doren, Vincent Lostanlen, Aurora Cramer, et al., "Automated acoustic monitoring captures timing and intensity of bird migration," *Journal of Applied Ecology* 60, no. 3 (2023): 433–444.

154   *bird migration patterns:* Frank A. La Sorte and Kyle G. Horton, "Seasonal variation in the effects of artificial light at night on the occurrence of nocturnally migrating birds in urban areas," *Environmental Pollution* 270 (2021): 116085.

154   *artificial light disorients:* Carrie Ann Adams, Esteban Fernández-Juricic, Erin Michael Bayne, and Colleen Cassady St Clair, "Effects of artificial light on bird movement and distribution: A systematic map," *Environmental Evidence* 10, no. 1 (2021): 1–28.

154   *one billion birds:* Scott R. Loss, Tom Will, Sara S. Loss, and Peter P. Marra, "Bird–building collisions in the United States: Estimates of annual mortality and species vulnerability," *The Condor* 116, no. 1 (2014): 8–23.

155   *BirdCast:* Brian L. Sullivan, Jocelyn L. Aycrigg, Jessie H. Barry, et al., "The eBird enterprise: An integrated approach to development and application of citizen science," *Biological Conservation* 169 (2014): 31–40.

CHAPTER 7: IMMERSION THERAPY

160   *healthcare and public policy:* Yuko Tsunetsugu, Bum-Jin Park, Juyoung Lee, et al., "Psychological relaxation effect of forest therapy: Results of field experiments in 19 forests in Japan involving 228 participants," *Nihon eiseigaku zasshi. Japanese Journal of Hygiene* 66, no. 4 (2011): 670–676.

160   *parasympathetic nervous system:* Jin Young Jeon, In Ok Kim, Poung-sik Yeon, and Won Sop Shin. "The physio-psychological effect of forest therapy programs on juvenile probationers," *International Journal of Environmental Research and Public Health* 18, no. 10 (2021): 5467.

160   *result from phytoncides:* Youngran Chae, Sunhee Lee, Youngmi Jo, et al., "The effects of forest therapy on immune function," *International Journal of Environmental Research and Public Health* 18, no. 16 (2021): 8440.

160   *improved creativity:* Eva M. Selhub and Alan C. Logan, *Your Brain on Nature: The Science of Nature's Influence on Your Health, Happiness and Vitality* (John Wiley & Sons, 2012).

160   *exposure to phytoncides:* Q. Li, M. Kobayashi, Y. Wakayaa, et al., "Effect of phytoncide from trees on human natural killer cell function," *International Journal of Immunopathology and Pharmacology.* 22, no. 4 (2009), 951–959.

161   shinrin-yoku, *or "forest bathing":* Margaret M. Hansen, Reo Jones, and Kirsten Tocchini, "Shinrin-yoku (forest bathing) and nature therapy: A state-of-the-art review," *International Journal of Environmental Research and Public Health* 14, no. 8 (2017): 851.

164   *three-quarters of Canadians:* Angela Eykelbosh and Anna Chow, "Canadian green spaces during COVID-19: Public health benefits and planning for resil-

ience," National Collaborating Centre for Environmental Health (NCCEH) Retrieved from https://ccnse. ca/sites/default/files/NCCEH% 20Canadian % 20greenspaces% 20and% 20COVID-19 (2022).

165 *park visitation:* Zoe M. Volenec, Joel O. Abraham, Alexander D. Becker, and Andy P. Dobson, "Public parks and the pandemic: How park usage has been affected by COVID-19 policies," *PloS One* 16, no. 5 (2021): e0251799.

169 *life indoors:* "Report to Congress on Indoor Air Quality: Volume 2. Assessment and Control of Indoor Air Pollution," Indoor Air Division, Office of Atmospheric and Indoor Air Programs, Office of Air and Radiation, and Office of Research and Development prepared by U.S. Environmental Protection Agency, (1989).

169 *vitamin D levels:* William B. Grant, Henry Lahore, Sharon L. McDonnell, et al., "Evidence that vitamin D supplementation could reduce risk of influenza and COVID-19 infections and deaths," *Nutrients* 12, no. 4 (2020): 988.

169 *lockdowns contributed to worsening mental health disorders:* Alexandra Burton, Alison McKinlay, Henry Aughterson, and Daisy Fancourt, "Impact of the COVID-19 pandemic on the mental health and well-being of adults with mental health conditions in the UK: A qualitative interview study," *Journal of Mental Health* (2021): 1–8.

181 *Text-A-Tree:* Julietta Sorensen Kass, Peter N. Duinker, Melanie Zurba, and Michael Smit, "Testing a novel human-nature connection model with Halifax's urban forest using a text-messaging engagement strategy," *Urban Forestry & Urban Greening* 65 (2021): 127350.

182 *"nature prescriptions":* Michelle C. Kondo, Kehinde O. Oyekanmi, Allison Gibson, et al., "Nature prescriptions for health: A review of evidence and research opportunities," *International Journal of Environmental Research and Public Health* 17, no. 12 (2020): 4213.

183 *scientific research proving the benefits of spending time in nature:* Phi-Yen Nguyen, Hania Rahimi-Ardabili, Xiaoqi Feng, and Thomas Astell-Burt, "Nature prescriptions: A scoping review with a nested meta-analysis," *medRxiv* (2022): 2022–03. https://www.medrxiv.org/content/medrxiv/early/2022/03/27 /2022.03.23.22272674.full.pdf.

183 *spending more time in nature:* Lotte Mortier De Borger, Hans Keune, and Lieve Peremans, "Natuur als therapie: hoe reageren patiënten op een 'groen voorschrift'?." *Huisarts nu: maandblad van de Wetenschappelijke Vereniging van Vlaamse Huisartsen.-Brussel* 51, no. 4 (2022): 192–195.

183 *"biophilia hypothesis":* E. O. Wilson, *Biophili.* (Cambridge: Harvard University Press, 1984).

183 *Stephen and Rachel Kaplan:* Rachel Kaplan and Stephen Kaplan, *The Experience of Nature: A Psychological Perspective* (Cambridge: Cambridge University Press, 1989).

184 *recovered from breast cancer:* Bernadine Cimprich, "Development of an intervention to restore attention in cancer patients," *Cancer nursing* 16, no. 2 (1993): 83–92.

184 *natural, therapeutic environments:* Bernadine Cimprich and David L Ronis,

"An environmental intervention to restore attention in women with newly diagnosed breast cancer," *Cancer Nursing* 26, no. 4 (2003): 284–292; quiz 293–24. doi:10.1097/00002820-200308000-00005.

184   *Roger Ulrich:* R. S. Ulrich, (1981). "Natural versus urban scenes: Some psychophysiological effects," *Environment and Behavior* 13, no. 5 (1981): 523–556.

184   *amygdala activity:* Sonja Sudimac, Vera Sale, and Simone Kühn, "How nature nurtures: Amygdala activity decreases as the result of a one-hour walk in nature," *Molecular Psychiatry* 27, no. 11 (2022): 4446–4452.

185   *complementary theories:* Nalise Hähn, Emmanuel Essah, and Tijana Blanusa, "Biophilic design and office planting: A case study of effects on perceived health, well-being and performance metrics in the workplace," *Intelligent Buildings International* 13, no. 4 (2021): 241–260.

186   *nature documentaries:* Florian Arendt and Jörg Matthes, "Nature documentaries, connectedness to nature, and pro-environmental behavior," *Environmental Communication* 10, no. 4 (2016): 453–472.

186   *prescribing the right exposure to nature:* Alex Hutchinson, "Nature Is Medicine. But What's the Right Dose?" *Outside Online*, September 14, 2021. https://www.outsideonline.com/health/wellness/naturequant-app-outdoor-data/.

186   *Park Rx America:* Robert Zarr, Linda Cottrell, and Chaya Merrill, "Park prescription (DC Park Rx): A new strategy to combat chronic disease in children," *Journal of Physical Activity and Health* 14, no. 1 (2017): 1–2.

186   *8,500 parks in the database:* Robert Zarr, "Nature Prescriptions: Renormalizing our relationship with the Great Outdoors," APHA's 2018 Annual Meeting & Expo (Nov. 10–Nov. 14).

187   *medicating using nature:* Robert Zarr, Bing Han, Erika Estrada, and Deborah A. Cohen, "The Park Rx trial to increase physical activity among low-income youth," *Contemporary Clinical Trials* 122 (2022): 106930.

188   *prescribing nature:* Melissa Sundermann, Deborah Chielli, and Susan Spell, "Nature as Medicine: The 7th (Unofficial) Pillar of Lifestyle Medicine," *American Journal of Lifestyle Medicine* (2023): 15598276231174863.

CHAPTER 8: INTO THE GREAT NEARBY

191   *De Nationale Bomenbank's fiftieth anniversary:* "Top 5," Nationale Bomen Top 50 (2022). https://www.denationalebomentop50.nl/nl/top5.

192   *benefits of nature extend far beyond easing our mood:* Terry Hartig, Richard Mitchell, Sjerp De Vries, and Howard Frumkin, "Nature and health," *Annual Review of Public Health* 35 (2014): 207–228.

192   *nature helps lengthen our life span:* David Rojas-Rueda, Mark J. Nieuwenhuijsen, Mireia Gascon, Daniela Perez-Leon, and Pierpaolo Mudu, "Green spaces and mortality: A systematic review and meta-analysis of cohort studies," *Lancet Planetary Health* 3, no. 11 (2019): e469–e477.

193   *life expectancy has stagnated:* "Life Expectancy in the US Dropped for the Second Year in a Row in 2021." *National Center for Health Statistics,* Centers for

Disease Control and Prevention. Retrieved from https://www.cdc. gov/nchs
/pressroom/nchs_press_releases/2022/20220831. htm#:~: text= Life% 20
expectancy% 20at% 20birth% 20in, its% 20lowest% 20level% 20since 201996
(2022).

193   *prevalence of chronic diseases:* Regina M. Benjamin, "Surgeon general's perspectives," *Public Health Reports* 128, no. 5 (2013): 350.

194   *shaping our lived experience:* Deirdre Pfeiffer and Scott Cloutier, "Planning for
happy neighborhoods," *Journal of the American Planning Association* 82, no. 3
(2016): 267–279.

194   *John Snow:* Narushige Shiode, Shino Shiode, Elodie Rod-Thatcher, Sanjay Rana,
and Peter Vinten-Johansen, "The mortality rates and the space-time patterns of
John Snow's cholera epidemic map," *International Journal of Health Geographics*
14, no. 1 (2015): 1–15.

200   *integral role that being in nature has on longevity:* Caoimhe Twohig-Bennett and
Andy Jones, "The health benefits of the great outdoors: A systematic review and
meta-analysis of greenspace exposure and health outcomes," *Environmental Research* 166 (2018): 628–637.

200   *promotes physical healing:* Julie H. Dean, Danielle F. Shanahan, Robert Bush,
et al., "Is nature relatedness associated with better mental and physical health?,"
International Journal of Environmental Research and Public Health 15, no. 7
(2018): 1371.

200   *zip code matters more than your genetic code:* Dannie Ritchie, "Our zip code may
be more important than our genetic code: Social determinants of health, law,
and policy," *Rhode Island Medical Journal* 96, no. 7 (2013): 14.

200   *average life expectancies vary by more than twenty years:* "Powerful Open Data
Tool Illustrates Life Expectancy Gaps Are Larger in More Racially Segregated
Cities," Build Healthy Places Network, July 8, 2019. https://www.buildhealthy
places.org/sharing-knowledge/blogs/expert-insights/powerful-open-data-tool
-illustrates-life-expectancy-gaps-are-larger-in-more-racially-segregated-cities/.

200   *disparities purely through social factors:* Elizabeth Arias, Loraine A. Escobedo,
Jocelyn Kennedy, Chunxia Fu, and Jodi A. Cisewski, "US small-area life expectancy estimates project: Methodology and results summary," National Center
for Health Statistics, Centers for Disease Control and Prevention, 2018.

200   *can help women live longer:* Peter James, Jaime E. Hart, Rachel F. Banay, and
Francine Laden, "Exposure to greenness and mortality in a nationwide pro-
spective cohort study of women," *Environmental Health Perspectives* 124, no. 9
(2016): 1344–1352.

201   *NatureDose:* Mondira Bardhan, Kuiran Zhang, Matthew HEM Browning, et
al., "Time in nature is associated with higher levels of positive mood: Evidence
from the 2023 NatureDose™ student survey," *Journal of Environmental Psychology* 90 (2023): 102083.

201   *U.S. life expectancy:* Sherry L. Murphy, J. Q. Xu, Kenneth D. Kochanek, and
Elizabeth Arias, "Mortality in the United States, 2017. NCHS data brief, No
328," *Hyattsville, MD: National Center for Health Statistics,* 2018.

201 *meta-analysis:* David Rojas-Rueda, Mark J. Nieuwenhuijsen, Mireia Gascon, Daniela Perez-Leon, and Pierpaolo Mudu, "Green spaces and mortality: A systematic review and meta-analysis of cohort studies," *Lancet Planetary Health* 3, no. 11 (2019): e469–e477.

201 *vegetation within 500 meters:* David Rojas-Rueda, Mark J. Nieuwenhuijsen, Mireia Gascon, Daniela Perez-Leon, and Pierpaolo Mudu, "Green spaces and mortality: A systematic review and meta-analysis of cohort studies," *Lancet Planetary Health* 3, no. 11 (2019): e469–e477.

202 *trees planted by Friends of Trees:* Geoffrey H. Donovan, Jeffrey P. Prestemon, Demetrios Gatziolis, et al., "The association between tree planting and mortality: A natural experiment and cost-benefit analysis," *Environment International* 170 (2022): 107609.

202 *email scams:* Kale Williams, "'More Trees, Fewer Deaths': Study Shows Life-saving Role of Portland's Tree Canopy," Kgw.com. December 14, 2022. https://www.kgw.com/article/tech/science/climate-change/portlands-study-livesaving-trees/283-cc6c8252-2c85-46e3-b566-050af19e7268.

205 *targeted investment in green infrastructure:* Jochem O. Klompmaker, Jaime E. Hart, Christopher R. Bailey, et al., "Racial, ethnic, and socioeconomic disparities in multiple measures of blue and green spaces in the United States," *Environmental Health Perspectives* 131, no. 1 (2023): 017007.

205 *NatureScore and COVID-19:* NatureQuant, July 6, 2020. https://www.naturequant.com/blog/Link-Between-Nature-and-Covid-19/.

205 *strong associations:* Jared Hanley, Christopher Minson, and Christopher Bailey, "Delivering Technology to Assess and Promote Nature Exposure," NatureQuant, 2020. https://www.naturequant.com/NatureQuant-whitepaper.pdf.

205 *City Builder:* "City Builder." n.d. City Builder. Accessed January 24, 2023. https://www.citivelocity.com/citybuilder/eppublic/cb/.

206 *"disproportionately upon underserved communities":* "Climate Change and Social Vulnerability in the United States: A Focus on Six Impacts," U.S. Environmental Protection Agency, 2021, EPA 430-R-21–003.

206 *NatureScore Priority Index (NPI):* Lorien Nesbitt and Jessica Quinton, "Invited Perspective: Nature Is Unfairly Distributed in the United States—But That's Only Part of the Global Green Equity Story," *Environmental Health Perspectives* 131, no. 1 (2023): 011301.

207 *report on the financial worth of nature:* Partha Dasgupta, "The Economics of Biodiversity: The Dasgupta Review," HM Treasury, UK, 2021. Accessed April 5, 2023. https://www.gov.uk/government/publications/final-report-the-economics-of-biodiversity-the-dasgupta-review.

208 *40 percent decrease in feelings of depression:* Eugenia C. South, Bernadette C. Hohl, Michelle C. Kondo, John M. MacDonald, and Charles C. Branas, "Effect of greening vacant land on mental health of community-dwelling adults: A cluster randomized trial," *JAMA Network Open* 1, no. 3 (2018): e180298–e180298.

208 *reductions in gun violence:* Charles C. Branas, Eugenia South, Michelle C. Kondo, et al., "Citywide cluster randomized trial to restore blighted vacant land

and its effects on violence, crime, and fear," *Proceedings of the National Academy of Sciences* 115, no. 12 (2018): 2946–2951.

209   *$15 million, five-year clinicial trial:* 2023. "Health, Environment and Action in Louisville (HEAL) Green Heart Louisville Project (HEAL)," U.S. National Library of Medicine, NIH, January 27, 2023. https://classic.clinicaltrials.gov/ct2/show/NCT03670524.

209   *Albert Lea, Minnesota:* Dan Buettner and Sam Skemp, "Blue zones: Lessons from the world's longest lived," *American Journal of Lifestyle Medicine* 10, no. 5 (2016): 318–321.

209   *The Blue Zones:* Dan Buettner, *The Blue Zones: Lessons for Living Longer From the People Who've Lived the Longest* (National Geographic Books, 2012).

210   *Blue Zones Vitality Project:* Hannah R. Marston, Kelly Niles-Yokum, and Paula Alexandra Silva, "A Commentary on Blue Zones®: A critical review of age-friendly environments in the 21st century and beyond," *International Journal of Environmental Research and Public Health* 18, no. 2 (2021): 837.

210   *life expectancy for the average participant rose by 3.2 years:* Amy Tomczyk, "Blue Zones Vitality City Project: Albert Lea, MN," Conduent, 2009. https://cdc.thehcn.net/promisepractice/index/view?pid=3911.

211   *Power Nine:* Lindsey V. Shelton, "How Community Engagement and the Power Nine Impact Health," Honors Theses, Murray State University, 2021.

211   *five Blue Zones:* Casandra Herbert, Mary House, Ryan Dietzman, et al., "Blue Zones: Centenarian Modes of Physical Activity: A Scoping Review," *Journal of Population Ageing* (2022): 1–37.

211   *the world's longest, healthiest life expectancies:* Jason Dean-Chen Yin and Alex Jingwei He, "Health insurance reforms in Singapore and Hong Kong: How the two ageing asian tigers respond to health financing challenges?," *Health Policy* 122, no. 7 (2018): 693–697.

211   *including Bangkok, Thailand:* "Thailand Data | World Health Organization," World Health Organization, Data.who.int. 2023. https://data.who.int/countries/764.

211   *green-space discrepancies:* "The Greening of Bangkok," United Nations Climate Change, Unfccc.int. March 21, 2021. https://unfccc.int/blog/the-greening-of-bangkok.

214   *partnership got immediate support:* Geoff Nudelman, "The Science of Tree Planting: How Data Is Boosting Climate Resilience of Both Cities and Forests," Sustainable Brands, July 17, 2023. https://sustainablebrands.com/read/product-service-design-innovation/science-tree-planting-data-boosting-climate-resilience-cities-forests.

215   *tree-planting campaign in Detroit:* Christine E. Carmichael and Maureen H. McDonough, "Community stories: Explaining resistance to street tree-planting programs in Detroit, Michigan, USA," *Society & Natural Resources* 32, no. 5 (2019): 588–605.

215   *"no-tree request":* Christine E. Carmichael and Maureen H. McDonough, "The trouble with trees? Social and political dynamics of street tree-planting efforts in Detroit, Michigan, USA," *Urban Forestry & Urban Greening* 31 (2018): 221–229.

215 *involvement in the decision-making:* Christine E. Carmichael, Cecilia Danks, and Christine Vatovec, "Green infrastructure solutions to health impacts of climate change: Perspectives of affected residents in Detroit, Michigan, USA." *Sustainability* 11, no. 20 (2019): 5688.

216 *Oregon's Urban and Community Forestry Program:* Adam Duvernay, "Eugene Forester Will Lead ODF Program Helping Cities Improve Their Urban Forests," *The Register-Guard.* July 23, 2022. https://eu.registerguard.com/story/news/2022/07/23/eugene-forester-scott-altenhoff-leads-oregon-department-forestry-program-cities-urban-forests/65361784007/.

218 *"Working with Vivek":* Sheraz Sadiq, "Using 'Green Infrastructure' to Promote Equity Is a Key Goal for New Oregon Forestry Manager," Oregon Public Broadcasting, August 17, 2022. https://www.opb.org/article/2022/08/05/using-green-infrastructure-to-promote-equity-is-a-key-goal-for-new-oregon-forestry-manager/.

220 *particularly useful in poor neighborhoods:* Susan Strife and Liam Downey, "Childhood development and access to nature: A new direction for environmental inequality research," *Organization & Environment* 22, no. 1 (2009): 99–122.

220 *access perspective:* Danielle F. Shanahan, B. B. Lin, K. J. Gaston, R. Bush, and R. A. Fuller, "Socio-economic inequalities in access to nature on public and private lands: A case study from Brisbane, Australia," *Landscape and Urban Planning* 130 (2014): 14–23.

CHAPTER 9: RAISING FUTURE NATURALISTS

224 *innate affinity for nature:* Camilla S. Rice and Julia C. Torquati, "Assessing connections between young children's affinity for nature and their experiences in natural outdoor settings in preschools," *Children Youth and Environments* 23, no. 2 (2013): 78–102.

226 *developing brains:* Hui Li, Wenwei Luo, and Huihua He, "Association of parental screen addiction with young children's screen addiction: A chain-mediating model," *International Journal of Environmental Research and Public Health* 19, no. 19 (2022): 12788.

226 *release of dopamine:* Gadi Lissak, "Adverse physiological and psychological effects of screen time on children and adolescents: Literature review and case study," *Environmental Research* 164 (2018): 149–157.

226 *child's cognitive development:* Helena Duch, Elisa M. Fisher, Ipek Ensari, and Alison Harrington, "Screen time use in children under 3 years old: A systematic review of correlates," *International Journal of Behavioral Nutrition and Physical Activity* 10 (2013): 1–10.

226 *emotional and social difficulties:* Gadi Lissak, "Adverse physiological and psychological effects of screen time on children and adolescents: Literature review and case study," *Environmental Research* 164 (2018): 149–157.

226 *prefrontal cortex:* Aric Sigman, "Screen Dependency Disorders: A new challenge

for child neurology," *Journal of the International Child Neurology Association* 1, no. 1 (2017).

226 *entree to outdoor exploration:* Maxine R. Crawford, Mark D. Holder, and Brian P. O'Connor, "Using mobile technology to engage children with nature," *Environment and Behavior* 49, no. 9 (2017): 959–984.

226 *eco-anxiety:* Charlie Kurth and Panu Pihkala, "Eco-anxiety: What it is and why it matters," *Frontiers in Psychology* 13 (2022): 981814.

226 *American Psychological Association:* Melody Schreiber, "Addressing Climate Change Concerns in Practice," *American Psychological Association* 51, no. 2 (2021), Apa.org. https://www.apa.org/monitor/2021/03/ce-climate-change.

227 *extremely concerned about climate change:* Caroline Hickman, Elizabeth Marks, Panu Pihkala, et al., "Climate anxiety in children and young people and their beliefs about government responses to climate change: A global survey," *Lancet Planetary Health* 5, no. 12 (2021): e863–e873.

227 *soothing the soul:* Robert Zarr, "Connecting People to Nature: The Evidence and Practice of Nature Prescriptions," APHA 2017 Annual Meeting & Expo (Nov. 4–Nov. 8).

227 *children and teenagers facing eco-anxiety:* Pauline Baudon and Liza Jachens, "A scoping review of interventions for the treatment of eco-anxiety," *International Journal of Environmental Research and Public Health* 18, no. 18 (2021): 9636.

228 *FatherLove:* Richard Louv, *Fatherlove: What We Need, What We Seek, What We Must Create* (Diane Publishing Company, 1993).

228 *Last Child in the Woods:* Richard Louv, *Last Child in the Woods: Saving Our Children from Nature-Deficit Disorder* (Algonquin Books, 2008).

228 *Nature Deficit Disorder:* Frances E. "Ming" Kuo, "Nature-deficit disorder: Evidence, dosage, and treatment," *Journal of Policy Research in Tourism, Leisure and Events* 5, no. 2 (2013): 172–186.

229 *nature schools:* Richard Louv, *Vitamin N: The Essential Guide to a Nature-Rich Life* (Algonquin Books, 2016).

229 *pediatricians:* Stephen J. Pont, Jaime Zaplatosch, Margaret Lamar, et al., "Green schoolyards support healthy bodies, minds and communities," *Pediatrics* 142 (2018): 733.

229 *movement continues to grow:* Cheryl Charles, Richard Louv, Lee Bodner, and Bill Guns, "Children and nature 2008," *A Report on the Movement to Reconnect Children to the Natural World. Santa Fe: Children and Nature Network* 9, no. 11 (2008).

229 *heralded as the pied piper:* Richard Louv, "Want Your Kids to Get into Harvard? Tell 'Em to Go Outside!," *New Nature Movement blog*, 2011.

230 *"anti-technology":* Cary Seidman, "Vitamin N," *Science Scope* 40, no. 1 (2016): 88.

230 *restricts little ones' freedom:* Jean S. Coffey and Lindsey Gauderer, "When pediatric primary care providers prescribe nature engagement at a State Park, do children 'fill' the prescription?," *Ecopsychology* 8, no. 4 (2016): 207–214.

231 *limiting children's nature exposure:* L. R. Larson, H. K. Cordell, C. J. Betz, and G. T. Green, "Children's time outdoors: Results from a national survey," *Proceedings of the Northeastern Recreation Research Symposium*, 2011.

231 *four hours per week:* "Children Spend Only Half as Much Time Playing Outside as Their Parents Did," *The Guardian,* July 27, 2016. https://www.theguardian .com/environment/2016/jul/27/children-spend-only-half-the-time-playing -outside-as-their-parents-did.

231 *outdoor play:* Julie Ernst, "Exploring Young Children's and Parents' Preferences for Outdoor Play Settings and Affinity toward Nature," *International Journal of Early Childhood Environmental Education* 5, no. 2 (2018): 30–45.

231 *spend less time outdoors than prison inmates:* Damian Carrington, "Three-Quarters of UK Children Spend Less Time Outdoors than Prison Inmates—Survey," *The Guardian*, March 25, 2016. https://www.theguardian.com /environment/2016/mar/25/three-quarters-of-uk-children-spend-less-time -outdoors-than-prison-inmates-survey.

231 *presents similarly to ADHD:* Benita R. Schmitz, "Nature-deficit disorder and the effects on ADHD," Seminar Paper, University of Wisconsin-Platteville, 2012.

231 *exposure to nature:* Frances E. Kuo and Andrea Faber Taylor, "A potential natural treatment for attention-deficit/hyperactivity disorder: Evidence from a national study," *American Journal of Public Health* 94, no. 9 (2004): 1580–1586.

231 *Danish study:* Malene Thygesen, Kristine Engemann, Gitte J. Holst, et al., "The association between residential green space in childhood and development of attention deficit hyperactivity disorder: A population-based cohort study," *Environmental Health Perspectives* 128, no. 12 (2020): 127011.

231 *6.1 million American children:* Melissa L. Danielson, Rebecca H. Bitsko, Reem M. Ghandour, et al., "Prevalence of Parent-Reported ADHD Diagnosis and Associated Treatment among U.S. Children and Adolescents, 2016," *Journal of Clinical Child and Adolescent Psychology* 47, no. 2 (2018): 199–212. Published online January 24, 2018.

232 *a fifth of U.S. children:* Adekunle Sanyaolu, Chuku Okorie, Xiaohua Qi, Jennifer Locke, and Saif Rehman, "Childhood and adolescent obesity in the United States: A public health concern," *Global Pediatric Health* 6 (2019): 2333794X19891305.

232 *are now obese:* Solveig A. Cunningham, Shakia T. Hardy, Rebecca Jones, et al., "Changes in the incidence of childhood obesity," *Pediatrics* 150, no. 2 (2022): e2021053708.

232 *myopia or nearsightedness:* Justin C. Sherwin, Mark H. Reacher, Ruth H. Keogh, et al., "The association between time spent outdoors and myopia in children and adolescents: A systematic review and meta-analysis," *Ophthalmology* 119, no. 10 (2012): 2141–2151.

232 *higher levels of outdoor activity and lower rates of myopia:* Kathryn A. Rose, Ian G. Morgan, Jenny Ip, et al., "Outdoor activity reduces the prevalence of myopia in children," *Ophthalmology* 115, no. 8 (2008): 1279–1285.

232 *outdoor play on myopia progression:* Pei-Chang Wu, Chia-Ling Tsai, Chia-Huo Hu, and Yi-Hsin Yang, "Effects of outdoor activities on myopia among rural school children in Taiwan," *Ophthalmic Epidemiology* 17, no. 5 (2010): 338–342.

232 *greater access to nature in children:* Louise Chawla, Kelly Keena, Illène Pevec,

and Emily Stanley, "Green schoolyards as havens from stress and resources for resilience in childhood and adolescence," *Health & Place* 28 (2014): 1–13.

232   *engaging sixth graders in three outdoor education programs:* Deborah Parrish, G. Phillips, R. Levine, et al., "Effects of outdoor education programs for children in California," *American Institutes for Research*, 2005.

232   *fostering ecological knowledge:* Matteo Giusti, Stephan Barthel, and Lars Marcus, "Nature routines and affinity with the biosphere: A case study of preschool children in Stockholm," *Children, Youth and Environments* 24, no. 3 (2014): 16–42.

233   *"technological nature":* Peter H. Kahn Jr., *Technological Nature: Adaptation and the Future of Human Life* (MIT Press, 2011).

233   *fail utterly as substitutes:* Peter H. Kahn Jr., Rachel L. Severson, and Jolina H. Ruckert, "The human relation with nature and technological nature," *Current Directions in Psychological Science* 18, no. 1 (2009): 37–42.

235   *Pokémon GO players:* Leejiah J. Dorward, John C. Mittermeier, Chris Sandbrook, and Fiona Spooner, "Pokémon Go: Benefits, costs, and lessons for the conservation movement," *Conservation Letters* 10, no. 1 (2017): 160–165.

235   *in light of the accidents:* Yvette Chong, Dean Krishen Sethi, Charmaine Hui Yun Loh, and Fatimah Lateef, "Going forward with pokemon go," *Journal of Emergencies, Trauma, and Shock* 11, no. 4 (2018): 243.

235   *Pokémon GO players suffered:* Stefania Barbieri, Gianna Vettore, Vincenzo Pietrantonio, et al., "Pedestrian inattention blindness while playing Pokémon Go as an emerging health-risk behavior: A case report," *Journal of Medical Internet Research* 19, no. 4 (2017): e86.

235   *colliding into trees:* Ali Pourmand, Kevin Lombardi, Evan Kuhl, and Francis O'Connell, "Videogame-related illness and injury: A review of the literature and predictions for Pokémon GO!," *Games for Health Journal* 6, no. 1 (2017): 9–18.

236   *MAR games help make children more social:* Prithwijit Das, Meng Zhu, Laura McLaughlin, Zaid Bilgrami, and Ruth L. Milanaik, "Augmented Reality Video Games: New Possibilities and Implications for Children and Adolescents," *Multimodal Technologies and Interaction* 1, no. 2 (2017): 8. Accessed July 25, 2023. https://doi.org/10.3390/mti1020008.

236   *opportunities of MAR games in environmental education:* A. Kamarainen, J. Reilly, S. Metcalf, T. Grotzer, and C. Dede, "Using mobile location-based augmented reality to support outdoor learning in undergraduate ecology and environmental science courses," *Bulletin of the Ecological Society of America* 99 no. 2 (2018): 259–276.

238   *SquirrelMapper project:* Jill Nugent, "Squirrel Mapper at a glance," *The Science Teacher* 87, no. 5 (2020): 17.

238   *helps researchers:* Miriam Brandt, Quentin Groom, Alexandra Magro, et al., "Promoting scientific literacy in evolution through citizen science," *Proceedings of the Royal Society B* 289, no. 1980 (2022): 20221077.

238   *squirrel observations:* Bradley J. Cosentino and James P. Gibbs, "Parallel evolution of urban–rural clines in melanism in a widespread mammal," *Scientific Reports* 12, no. 1 (2022): 1752.

239   *nearly 1.9 million organisms:* "2023 Results," City Nature Challenge, 2023. https://www.citynaturechallenge.org/current-results.

239   *a sign of hope for conservationists:* V. Cambria, F. Buffi, F. Attorre, et al., "City Nature Challenge: An effective Citizen Science approach for monitoring urban biodiversity," *Conferenza Annuale di LifeWatch Italia-Roma, 25–27 giugno 2018-Abstract Book*, pp. 43, IT, 2018.

239   *"white-spotted slimy salamander"*: Ann Cameron Siegal, "City Nature Challenge Lets Kids Help Scientists around the World," *Washington Post*, April 24, 2022. https://www.washingtonpost.com/kidspost/2022/04/24/city-nature-challenge-citizen-science/.

240   *been recorded in the county since 1977:* Alonso Abugattas, "Arlington's Natural Treasures and Protection Efforts," 2020. https://www.arlingtonva.us/files/sharedassets/public/environment/documents/arlingtons-natural-treasures-and-protection-efforts.pdf.

240   *young stewards:* Colin Campbell, "'It Was Fun, but There Were a Lot of Bugs': Citizen Scientists Document Wildlife at Maryland Zoo for Earth Day," *Baltimore Sun,* April 21, 2019. https://www.baltimoresun.com/science/bs-md-citizen-science-earth-day-20190421-story.html.

240   *a new generation with the wonders of the outdoors:* Soledad Altrudi, "Connecting to nature through tech? The case of the iNaturalist app." *Convergence* 27, no. 1 (2021): 124–141.

240   *inform local land management decisions:* Estibaliz Palma, Luis Mata, Kylie Cohen, et al., "The City Nature Challenge—A global citizen science phenomenon contributing to biodiversity knowledge and informing local government practices," *bioRxiv* (2022): 2022–11.

241   *La Paz:* "La Paz Is Declared Winner of the 2023 City Nature Challenge, Surpassing 450 Cities around the World," Sdsn Bolivia, May 23, 2023. https://sdsnbolivia.org/en/la-paz-is-declared-winner-of-the-2023-city-nature-challenge-surpassing-450-cities-around-the-world/.

241   *train the students:* Stephen Sautner, "La Paz, Bolivia Breaks Records in City Nature Challenge 2022," Newsroom.wcs.org, May 10, 2022. https://newsroom.wcs.org/News-Releases/articleType/ArticleView/articleId/17515/La-Paz-Bolivia-Breaks-Records-in-City-Nature-Challenge-2022.aspx.

241   Phylodryas boliviana *snake*: "Philodryas Boliviana," Reptile Database, n.d. Accessed July 25, 2023. https://reptile-database.reptarium.cz/species?genus=Philodryas&species=boliviana.

243   *training fledgling pilots:* Richard Louv, *The Nature Principle: Reconnecting with Life in a Virtual Age* (Algonquin Books, 2012).

244   *"dumbest generation"*: M. Bauerlein and S. G. Walesh, *The Dumbest Generation: How the Digital Age Stupefies Young Americans and Jeopardizes Our Future* (Jeremy P. Tarcher/Penguin, New York, 2009); ISBN 978–1–58542–639–3.

244   *"hybrid mind"*: Richard Louv, "The Hybrid Mind: The More High-Tech Schools Become, the More Nature They Need," *New Nature Movement blog, Children & Nature Network, November* 18, 2013.

INDEX

ABOUT

MARINER BOOKS

MARINER BOOKS traces its beginnings to 1832 when William Ticknor co-founded the Old Corner Bookstore in Boston, from which he would run the legendary firm Ticknor and Fields, publisher of Ralph Waldo Emerson, Harriet Beecher Stowe, Nathaniel Hawthorne, and Henry David Thoreau. Following Ticknor's death, Henry Oscar Houghton acquired Ticknor and Fields and, in 1880, formed Houghton Mifflin, which later merged with venerable Harcourt Publishing to form Houghton Mifflin Harcourt. HarperCollins purchased HMH's trade publishing business in 2021 and reestablished their storied lists and editorial team under the name Mariner Books.

Uniting the legacies of Houghton Mifflin, Harcourt Brace, and Ticknor and Fields, Mariner Books continues one of the great traditions in American bookselling. Our imprints have introduced an incomparable roster of enduring classics, including Hawthorne's *The Scarlet Letter,* Thoreau's *Walden,* Willa Cather's *O Pioneers!,* Virginia Woolf's *To the Lighthouse,* W.E.B. Du Bois's *Black Reconstruction,* J.R.R. Tolkien's *The Lord of the Rings,* Carson McCullers's *The Heart Is a Lonely Hunter,* Ann Petry's *The Narrows,* George Orwell's *Animal Farm* and *Nineteen Eighty-Four,* Rachel Carson's *Silent Spring,* Margaret Walker's *Jubilee,* Italo Calvino's *Invisible Cities,* Alice Walker's *The Color Purple,* Margaret Atwood's *The Handmaid's Tale,* Tim O'Brien's *The Things They Carried,* Philip Roth's *The Plot Against America,* Jhumpa Lahiri's *Interpreter of Maladies,* and many others. Today Mariner Books remains proudly committed to the craft of fine publishing established nearly two centuries ago at the Old Corner Bookstore.